Fukushima Meltdown

The World's First Earthquake-Tsunami-Nuclear Disaster

Takashi Hirose

This book is a translation of *Fukushima Genpatsu Merutodaun (Fukushima Nuclear Power Plant Meltdown)* by Takashi Hirose.

It was originally published as an Asahi Shinsho book (no. 298), by Asahi Shimbun Publications, Inc. on 30 May, 2011. The ISBN of the Japanese original is 978-4-02-273398-6. This translation is published with the permission of both the Author and Asahi Shimbun Publications, Inc.

Cover design   Morisaki Tadashi

The translation team consisted of Izumi Kaori, Kawauchi Alice, C. Douglas Lummis, Matt Kazuko, Niwa Shinobu, Okuda Hiroyuki, Ono M. Yoko, Otsuka Teruyo, and Yoko from LA

Fukushima Meltdown
The World's First Earthquake-Tsunami-Nuclear Disaster

Table of Contents

Concluding Chapter: Japan's Nuclear Power Policy is Completely Bankrupt

Is Nuclear Power the Trump Card Against Global Warming?

Without Nuclear Power, Will There Be Blackouts?

Japan's Nuclear Power Plants at a Dead End

The Final Disposition of Meltdown

Translator's Foreword

Or to be accurate, this should be called, One-of-the-translators' foreword. This book was translated by a team, of which I was one member. The work was divided among us. My share was to translate the introductory and final chapters and, as a native English speaker, to proofread and unify the style in the other chapters.

The team was motivated by the same sense of urgency which the author felt in writing it. After the combined earthquake-tsunami-nuclear disaster of 3/11 much of the country – especially in regions that seem "comfortably" far from Fukushima – is refusing to grasp the import of what has happened, and falling back into complacency. And this is giving people outside Japan an excuse to do the same. We hope that the English translation of Hirose Takashi's book – which he wrote in an explosion of energy in a matter of weeks after the disaster – will help people around the world come to a more accurate understanding of what is happening in this country. Hirose Takashi has been writing books against nuclear power since the early 1980s, one of the first being Tokyo ni, Genpatsu wo! (Nuclear Plants in Tokyo! [1981], a marvelous satire in which he argued, if these plants are as safe as the government says, then why not build them in downtown Tokyo, rather than in faraway places from which, when you send back the electricity, half of it is lost heating the wires. This critique of big-city egoism applies today: the reader will surely have noticed that the nuclear reactors that are melting down in Fukushima today are owned by Tokyo Electric Power Company, and the electric power they (used to) generate is delivered to Tokyo. They are in Fukushima because the people in Tokyo, while they wanted the electricity, didn't want the danger. The people of Fukushima and its neighboring regions get their electricity from a different company, Tohoku Electric.

Since then he has written several score more books, the most recent (before this one) being Genshiro Jigen Bakudan (Nuclear Reactor Time Bomb, Diamond, 2010). His work has been criticized for always imagining the worst case. Now that the worst case has come to pass, this criticism no longer can be used. In insisting that we consider the worst case he is no different, for example, from the fire department, whose job it is, always, to think, What if there's a fire?

Concerning the translation, some explanations are in order. As Hirose explains in his Introductory Chapter, some years ago the seismologist Ishibashi Katsuhiko coined a term for a new kind of catastrophe that had never yet happened, but could. The term, in Japanese, is genpatsu shinsai. What it means is a combined earthquake-nuclear disaster. Gen means "nuclear" patsu is an abbreviation for "power plant", shin is the short form for "earthquake" and sai means "disaster." The tragic fact is that Ishibashi's fear is now a reality: The Fukushima Disaster is the first time this phenomenon has been seen on this earth. The horror is something unprecedented. If what had happened were only an earthquake and tsunami, the destruction would be over now, and the survivors would long ago have been able to return to their homes, or their property if their homes were ruined, pick through the rubble to find what could be found, clean up, begin rebuilding. The whole region would be a great bustle of cleanup and reconstruction, filled with the sounds of hammers and power tools, trucks going in and out. But to the worst hit areas survivors can't go back and cleanup crews and construction workers can't get in. To the more outlying areas they shouldn't go in, but many do, at risk of their lives. No one knows how great the risk is. Neither the government nor the electric companies, nor their mouthpiece "experts", are being honest about what the risk is. And the reactors are, so far as we know, still leaking. In short, this is called the 3/11 Disaster by many, but it did not happen on 3/11, it began on 3/11, and is continuing today.

To return to the translation problem, as the phenomenon is unprecedented, there is no name for it in English. If we translate genpatsu shinsai literally, it would have to be something like "nuclear power plant earthquake disaster," pretty awkward, given that it appears repeatedly in the text. After much consultation among the translation team, and with the author (who is himself a skilled translator), we decided to use the Japanese term. Japanese words don't often find their way into English, but some do, tsunami being one of them. And it turns out that we are not the first to introduce genpatsu shinsai into English: Professor Ishibashi, who coined the term in 1997, has been doing so at least as far back as 2003. (1) We decided to add "syndrome" to the expression, to emphasize the fact that the disastrous effects of the earthquake, the tsunami, and the nuclear power plant accident are not simply added on to one another, but react on one another to produce results that none of them would produce by itself: the sum is greater than its parts. Thus in the text below, genpatsu shinsai syndrome is used as an English term of foreign origin, and thus is written without italics.

Concerning the names of earthquake disasters, there is a difference between Japanese and English customary usage. In English, the San Francisco Earthquake is used both to mean the seismic event, and the disaster that followed. In Japanese, these are generally separately described. The seismic event will be called jishin – earthquake, and the disaster will be called shinsai – earthquake disaster. Thus in the case of the 3/11 earthquake, the seismic event is officially called tohoku chihou taiheiyou oki jishin (rendered into English by Japan's Weather Bureau as "The 2011 off the Pacific Coast of Tohoku Earthquake"), whereas the disaster it caused is called higashi nihon daishinsai (literally, Great Eastern Japan Earthquake Disaster). Both of these are pretty long and awkward in English. In this text we will, for the seismic event, adopt the name given it by the US Geological Survey: the Great Tohoku Earthquake ("Tohoku" meaning "northeast"). And for the effect of that event, we will adopt a similar usage: the Great Tohoku Disaster.

Following East Asian practice, Japanese names are given with the family name first, with the single exception of the author's name where it appears on the cover and title page. The reason for the exception is that if you put the author's name on the cover in the Japanese order, the book will be filed in catalogues, libraries and bibliographies under the author's given name, and so get lost. Thus, the author of this book is Hirose Takashi; on the cover and title page he will be identified as Takashi Hirose. The team that produced the English version of this book consisted of Izumi Kaori, Kawauchi Alice, Matt Kazuko, Niwa Shinobu, Okuda Hiroyuki, Ono M. Yoko, Otsuka Teruyo, Yoko from LA and

C. Douglas Lummis

(1) See the prophetic, Ishibashi Kazuhiko, "Genpatsu-Shinsai: Multiple Disaster of Earthquake and Quake-induced Nuclear Accident Anticipated in the Japanese Islands."

Presented at the 23rd. General Assembly of IUGG (the International Union of Geodesy and Geophysics), 2003, Sapporo, Japan.

Author's Introduction to the English Edition

This book was written in the state of emergency created by the Fukushima Meltdown Accident, and takes into account only the events that took place in the first month and a half after that. Important things have happened since then. Among them is the decision by Prime Minister Kan Naoto, on 6 May, to require Chubu Electric to shut down (as explained in these pages, Japan's most earthquake-threatened reactor) the Hamaoka Nuclear Power Plant. And on 14 May Hamaoka was in fact taken off line. However this is only a temporary measure, which does not mean the plant is to be dismantled, so the earthquake danger remains as before.

Then in June, the Nuclear and Industrial Safety Agency (NISA, i.e. the cops) got together with Tokyo Electric Power Company (TEPCO, i.e. the robbers) and produced a report for the International Atomic Energy Agency (IAEA) which gave the tsunami as the sole cause for the accident. This was a falsified "simulation" of the accident, which concealed the fact that pipes were fractured by the earthquake before the tsunami came. With fearsome radiation pollution still continuing, Japan's Nuclear Power Mafia repent of nothing, and the country's state of danger continues.

But Japan is not the only country in danger. On 27 April this year, when in the US southern Alabama and other regions were attacked by tornadoes that killed more than 300 people, the Brown's Ferry Nuclear Power Plant's #3 reactor lost its external power supply and went into emergency shutdown and a critical state; a major accident was averted only by the electricity provided by an emergency diesel generator. Again, because of melting snow and heavy rain this May the river began to rise around the pressurized water reactor at Fort Calhoun-1 in Nebraska. On 17 June it began spilling over the levee, on 26 June it washed away the emergency levee, surrounded the high ground the plant is on, and inundated the plant. For a period the reactor lost its outside source of power, the cooling system for the nuclear fuel pool shut down, and again the situation was saved by the emergency diesel generator. And on 26 June a wildfire broke out near New Mexico's Los Alamos National Laboratory, which spread into the grounds of the Laboratory on the 27th. This was extinguished, but after that the main fire continued for a long period, and at the time of this writing the Laboratory is preparing for the flooding that typically follows fires in New Mexico.

Germany, Switzerland, Italy one after another have started moving rapidly towards the abolition of nuclear power. Even in Nuclear Great Power France, 77% now oppose nuclear power. Anywhere in the world, nuclear power plants are a wildly dangerous way to get electricity, and are unnecessary. The world needs to learn quickly from Japan's tragedy.

11 July, 2011   Hirose Takashi

Introduction: Nuclear Disaster Continues to Threaten

I want to write this book without resorting either to difficult technical language, or to evasive expressions like "beyond expectation" or "no immediate harmful effects." I will try to translate the opaque language of the "experts" who speak on television and write in the newspapers, and to focus on simple facts and short-term predictions regarding the relationship between nuclear power plants and earthquakes. But as I do this, the reader may find that it is mostly things that he or she had not known – had not been informed of.

The Fukushima Daiichi Nuclear Power Plant Disaster is a Human-made Disaster

On 11 March, 2011 at 2:46 PM, 130km off the Ojika Peninsula in Miyagi Prefecture, there occured a massive earthquake centering on the Pacific Plate. This was the Great Tohoku Earthquake. This earthquake and the ensuing tsunami brought destruction to the entire northeastern coast all the way to the Kanto [Tokyo] area, and became the Great Tohoku Disaster. Moreover, it caused damage so far as to cause a meltdown (I will explain what this is in Chapter 1) at Fukushima Daiichi Nuclear Power Plant.

However these two, the physical damage brought by the catastrophe and the nuclear power plant accident – though both were brought about by the same earthquake and seem like two aspects of a single event – are not the same, and a clear distinction must be made between them. For in their essential nature, they are different.

Natural disasters will never disappear. The Tokyo University seismologist Terada Torahiko left us with the epigram, "Natural disasters always come when we have forgotten about them", and we live in a time when we have forgotten his epigram.

The victims of the Great Tohoku Disaster have lost people dear to them, and seen the homes in which their lives were carried on, washed away. However these people do not raise their voices in complaint against anyone, but bear it silently. When one thinks of the enormity of their loss, it pains the heart, but the fact that there is no way of putting an end to earthquakes and tsunamis is, while something to which we cannot consent, is still something that we must accept. That is the fate of those of us who live on the Japanese Archipelago.

However the Fukushima Daiichi Nuclear Power Plant Disaster is neither a natural disaster nor ordained by fate. It is a human-made disaster brought about by bad faith. Unlike natural disasters, which are beyond human understanding, this was easily predictable and preventable. The officials of the Tokyo Electric Power Company (TEPCO) say things like, "We could not imagine that a once-in-a-thousand-years earthquake might come," and "The tsunami was beyond our expectation", but these are only obfuscations. A major earthquake accompanied by a nuclear power plant accident was – as I will show in detail – something very much within the realm of possibility, a possibility these officials arrogantly refused to consider.

The easy use by TEPCO and the Government of the expression "beyond expectation" cannot be allowed. The Fukushima Daiichi Nuclear Power Plant accident could have been prevented. That the victims desperately trying to hold their ground against the natural disaster should on top of that be showered by nuclear radiation caused by a ridiculous human-caused disaster fills me with both deep sadness and fierce anger.

My Gloomy Prediction Came True

The seismologist Ishibashi Katsuhiko (presently Emeritus Professor at Kobe University) predicted that a nuclear power plant accident like the present one was possible, and issued warnings from the late '90s. People had been warning of the danger of earthquake-caused nuclear accidents since the 1970s, but Ishibashi, from the specialized standpoint of seismology, proposed a new concept, which he called genpatsu shinsai [Translator's note: this expression literally means Nuclear-Power-Plant-Earthquake Disaster. As there is no English expression for this (the phenomenon itself is new) in this work we will render it as genpatsu shinsai syndrome. See the Translators' Foreword.]. By this he meant a situation in which, as the damage from the earthquake widens, the situation is made doubly worse by nuclear radiation damage. It was in the hope of preventing this that he was issuing warnings. Ishibashi wrote many books on this, including *Daichidouran no Jidai* (*The Age of Shifting Earth*) (Iwanami Shinsho) and he is a very well-known scholar, so it is impossible that his warnings were unknown to the officials of TEPCO.

I myself published in August of last year (2010) Genshiro Jigen Bakudan – Daijishin ni Obieru Nihonretto (Nuclear Reactor Time Bomb – The Japanese Archipelago under the Danger of Great Earthquakes) (Diamond) in which I tried to give the warnings of Ishibashi and others a wider audience, and to call for preventive measures to be taken against a genpatsu shinsai syndrome. Though as for earthquakes and tsunamis I am no more than a layman, I forecast the possibility of such a catastrophe, and wrote about nuclear contamination. It is a terrifying subject. Nevertheless, on the possibility of a genpatsu shinsai syndrome I wrote as follows.

If you ask, "Will Japan still be here in ten years?" I have the evil foreboding that the answer might be, "There is a very strong chance that it will not" . . . . In the future there awaits an unknowable, vast dark age. I don't want to contemplate its form, but it is the fear of a genpatsu shinsai syndrome brought about by movements of the earth that no human knowledge can control.

Today this evil foreboding has become a reality, and is getting worse day by day. If I, neither a scholar nor a specialist, was able to foresee this, and the nuclear power specialists from TEPCO and from the government's nuclear-related agencies were not, then for what do they exist?

A Mass Media that Reports Neither Facts nor Predictions

When I finished writing Nuclear Power Plant Time Bomb, I was hoping deep inside my assertion that a genpatsu shinsai syndrome was possible would prove to be a mistake. I was praying that nuclear power would be abolished, and the things that I had written would not come to pass. At the same time I resolved within myself that if things did occur as I had described, I would say nothing further.

I know many people who were victims of various nuclear power plant disasters, beginning with the 1986 accident in the #4 reactor at Chernobyl in the then Soviet Union (now Ukraine). I thought, if the sense of crisis I have put in my books fails to reach the public, and the same kind of victimization begins to appear among the people near to me, my sadness and regret will have no end, and my mortification at my powerlessness will be unbearable. If that happened, all I could hope for would be that as many people as possible were saved. Because I would not have the courage to turn my eyes toward the victims.

However, even though now that the genpatsu shinsai syndrome has actually occurred, the people of Japan seem not to have grasped the nature of the crisis. Unbelievably, though anyone who looks can see that the situation is getting worse day by day, television and the other media keep repeating "there is no crisis", "There's nothing to worry about" over and over, and show no inclination to report the critical nature of what is happening at Fukushima Daiichi. The obfuscations such as those I mentioned above, "The tsunami was beyond our expectation", "Such small amounts of radiation have no effect on health", and other statements by government people that obscure the dangers of exposure to radiation step outside the realm of the permissible.

In the past, on television only scholars who are proponents of nuclear power have given commentary as "experts". We listen to them telling their fantastic lies and offering their impossibly optimistic predictions, and as a result the people, having been given none of the facts, have gone on living as normal right up to the edge of the collapse. Now everything the "experts" told us has proved wrong, and the worst has happened. What is remarkable is that these people who failed to predict the genpatsu shinsai syndrome are shameless enough to put on their "expert" masks, reappear on television, and give their commentaries on the accident. The people responsible for the horror of this nuclear accident are the people who promoted nuclear power. The people in the TV stations who brought in this criminal gang and put them on the air day after day are equally responsible. Has there ever been even once that a television station warned the public that a nuclear power plant might be destroyed by an earthquake? Or a station that warned of the danger of a tsunami? After it has happened, even a child can do it.

Their ignorance and incompetence is proved day by day. I will treat this in more detail in Chapter 1, but on 12 March, 2011, when the No. 2 reactor at Fukushima Daiichi underwent a hydrogen explosion and it was clear that the worst had happened and a large amount of radioactive material had escaped outside, the commentators were at pains to persuade us that the crisis was very slight. The mainstream media not only did not warn the people about the danger of a genpatsu shinsai syndrome before it happened, but even after it had, went on broadcasting as if it hadn't. This is going on right now. This is as terrifying as what is happening at the actual site.

Then a month after the accident, on 11 April, the Nuclear Safety Commission (NSC, the Government agency that oversees the nuclear power industry) announced that the Fukushima Daiichi Power Plant "in the first hours after the accident, was emitting as much as 10,000 terabecquerels of radiation per hour (one terabcquerel = one trillon becquerels – bq)", and the Nuclear and Industrial Safety Agency (NISA, an agency of Ministry of Economy, Trade and Industry – METI) acknowledged that the accident was at the same level as Chernobyl. According to data taken between 11 March, the day of the disaster, and April 5 the total amount of radioactive iodine 131 and cesium 137 released into the atmosphere, as of 23 March, was more than 100,000 terabecquerels, several tens of thousands of terabecquerels above the standard for a Level 7 accident. From the first days after the earthquake I had been saying that huge amounts of radiation were sure to be escaping, but people only pointed their fingers at me as a rumor monger. A month later, the government acknowledged that the "rumor" was true. From day one the situation had reached the highest level for nuclear accidents, Level 7, and from day one the government knew this, but it concealed that information from the people, thus causing far more people to be irradiated than otherwise would have been the case. Or if they were unable to know this from the beginning, this only proves that they have neither competence at measurement nor sound judgment, which is more serious still.

Next, the television stations that announced the news about Level 7 all followed that with repetitions of the NISA's statement that still, "the amount of radiation released at Fukushima was only one tenth of that released at Chernobyl".

If Fukushima released one tenth of the radiation released at Chernobyl, what would happen? How many people know how many radiation victims there have been in the three countries that resulted after the demise of the Soviet Union – Russia, the Ukraine, and Belarus? In 2004, 18 years after the accident, the Welfare Ministry of the Ukraine released documents indicating that the Chernobyl accident produced 3,200,000 radiation victims, and that 2,300,000, including 450,000 children, were under government care. The following year the Russian Ministry of Healthcare and Social Development announced that there were in Russia more than 1,450,000 people whose health had been damaged by the Chernobyl accident and 226,000 children born after the accident, and thus under 18, whose health had been damaged by the accident. Among the victims, 46,000 had been certified as physically disabled. In March, 2006, 20 years after the accident, the people whose health had been damaged in Russia, the Ukraine and Belarus numbered 7,000,000. On 26 April, 2009 a memorial was held for the Chernobyl victims in the Ukraine, where more than 2,300,000 people have been officially recognized as radiation victims, and 4,400 who were children or youths at the time of the accident, were irradiated by radioactive iodine, and have been operated on for thyroid cancer. From this we can understand what is going to happen in Japan, whose population density is far greater than that of Russia, the Ukraine and Belarus.

It is people like these, who are unaware of these historical facts, and who remain unabashedly ignorant of the dangers of radiation, who control Japan's mass media. To say that "the radiation released is only one tenth that of Chernobyl", which means that hundreds of thousands of people have been exposed to serious levels of radiation, is to take the shameful position that you are not interested in learning what is going to happen, but only in protecting yourself. This is not easy to forgive.

In the event, I find I am not able to remain silent. When Asahi Shinsho contacted me, I decided that I did have to write one more book. I realized that, to try to save our children and young people, I must make known to as many people as possible, and as soon as possible, the things that the television "experts" refuse to deal with, even though the crisis is upon us: the essential, common-sense facts about nuclear power plants, and the likely form that the next genpatsu shinsai syndrome will take.

If we do not understand the seriousness of the Fukushima Daiichi Power Plant accident, we will certainly be hit by the #2 and #3 genpatsu shinsai syndrome before long. I am positive of this. We absolutely must prevent that from happening. If a genpatsu shinsai syndrome surpassing Fukushima does occur – then I will surely fall silent. I will write no more books about nuclear power. Writing such books will no longer have any meaning.

Can A Genpatsu Shinsai Syndrome at Hamaoka Be Prevented?

Today, 27 April, 2011, as I finish up the writing of this book, TEPCO is pumping water into the #1-4 reactors at Fukushima Daiichi, trying to cool them down. Today, too, large amounts of highly irradiated water from the reactor pressure vessels and the containment vessels continue to leak outside.

How will this accident end? Because TEPCO is concealing the danger from us, I imagine the worst. Suppose (though this is almost impossible) that everything came together perfectly: in each reactor the electrical circuits were completely restored, the pumps began circulating cooling water, and the temperature inside was reduced to below 100 degrees centigrade (cold shutdown). But large amounts of radioactive materials have already escaped and have been spread over a wide area; the sea to the south has already been terribly polluted. And it is certain that radioactivity is leaking out the bottom of the reactors.

At this point, rather than hoping that no greater amount of radioactivity will be released, what is important is for people from Fukushima to as far south as Tokyo – especially children and pregnant mothers – judging on the basis of the objective fact that radiation is, now, being released, insofar as they can, to take action to escape from irradiation. I say this because, the foolish soothing words of the television announcers that "the radiation levels are going down" to the contrary notwithstanding, the fact is that in the Japanese archipelago radiation pollution is accumulating day by day. Escaped radiation decreases almost not at all. What is frightening is that this radiation, tasteless and odorless, dribbling out and collecting year after year, will work its way into the bodies of children, and eat away humans and other living things from the inside. And the enrichment of radiation still continues.

And aside from escaping the radiation, the people in this country must quickly begin action to prevent the next genpatsu shinsai syndrome. How many members of the public know how many nuclear reactors there are in this country? From Hokkaido's Tomari Village to Kagoshima Prefecture's Satsuma-Sendai city in the south, in 17 locations there are, including those at Fukushima, 54 reactors in commercial operation. In addition, in Fukui Prefecture's Tsuruga Town there is a fast-breeding reactor, and in Aomori Prefecture's Rokkasho Village there is a nuclear fuel reprocessing facility. It has all the aspect of a Nuclear Power Archipelago, but if you examine these facilities carefully, you will understand that each of them, if hit by an earthquake on the same scale as that which hit Fukushima, could bring catastrophe.

I will discuss this in more detail in Chapter 4, but in January, 1995, beginning with the Southern Hyogo Prefectural Earthquake that brought about the Great Hanshin Earthquake Disaster, the Japanese Archipelago entered a period of frequent earthquakes. Following the 3/11 Great Tohoku Earthquake, which was a plate boundary earthquake, it is very likely that a much larger earthquake will occur on an active fault directly underground. If such an earthquake were to occur directly below an operating nuclear power plant, what would happen?

The greatest present danger of another genpatsu shinsai syndrome is the impending Tokai Earthquake, whose expected epicenter is directly beneath Chubu Electric's Hamaoka Nuclear Power Plant (Omaezaki City, Shizuoka Pref.). That "The likelihood of the Tokai Earthquake occurring within the next 30 years is 87%" is the judgment of the government's Headquarters for Earthquake Research Promotion. Whether in thirty years or tomorrow, this earthquake is sure to come. Hamaoka is near the huge metropolitan areas of Tokyo and Nagoya, and is adjacent to both the Tokaido Shinkansen (Bullet Train) line and the Tomei Expressway. In this book I want to argue that all of Japan's 54 nuclear power plants should be shut down, but as the first priority, the Hamaoka Plant should be shut down immediately. In this area where the population density is far greater than that of Fukushima, an even larger earthquake is threatening: this is the danger from which we must not avert our eyes.

The Wolf Will Come

By warning of a second and even third genpatsu shinsai syndrome following Fukushima, perhaps people will say, "What, are you trying to create panic?" But panic is what occurs when people don't know the truth.

If you tell people vague things such as that radiation "will have no immediate harmful effects", when the radiation does begin to take effect, that is when panic breaks out. Only when people know the true dangers can they respond swiftly and with composure. In this book, I will write what I know about the dangers of genpatsu shinsai syndrome. The decision as to whether, using this as a basis for judgment, to take action can only be made by the reader.

Among Aesop's Fables, there is the story, "The Boy Who Cried Wolf". A young shepherd boy who is bored with his work cries, "A wolf is coming! A wolf is coming!" The adults come running, but the wolf never appears. Thinking it fun, the boy cries out again, the adults come running, but again there is no wolf.

But when the wolf finally does come, and the boy cries "He's here!" and pleads for help, no one comes and the sheep are eaten.

Think of me as The Boy Who Cried Wolf. In this book there are no lies. The wolf might not come tomorrow. But that does not mean that the wolf is not coming, only that it may take a little more time. In the end, the wolf will come. When that happens, unless we take preventive measures in advance, it will not be simply a matter of some sheep; our very lives will be destroyed.

Hirose Takashi

Chapter 1:
A Human-made Disaster Triggered by a Tsunami

Reactors #1 through #6 of TEPCO's Fukushima Daiichii Nuclear Power
Plant lay side by side along the coast between the towns of Futabacho and
Ohkumamachi in Fukushima Prefecture.

It was 26 March 1971 when #1 reactor began operation. And it was just
two weeks before the 40th anniversary of that date that the Great Tohoku
Earthquake hit the prefecture, causing a genpatsu shinsai syndrome.

A nuclear power plant is intricately constructed with multiple parts and
components. The lifespan of each part is around 30 years at maximum.

Figure 1
Years when the Fukushima Plants started and their output

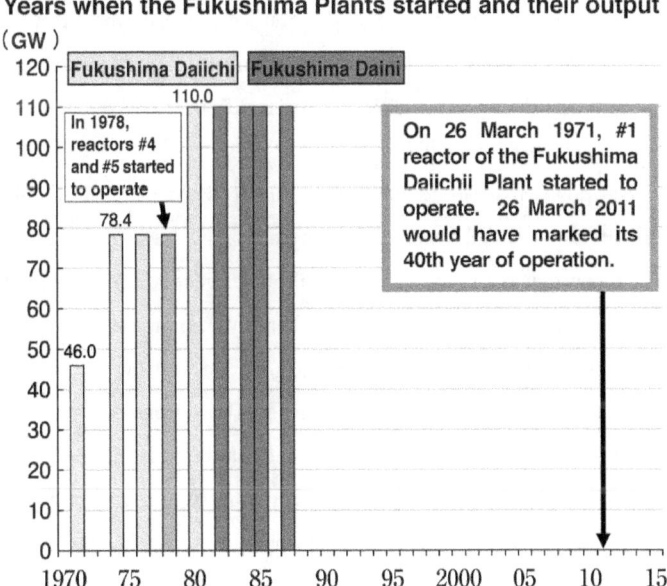

This 30 year lifespan applies not only to parts of reactors but also to parts of other machines such as elevators and automobiles.

The #1 reactor at the Fukushima Daiichii Plant, after 40 years in operation, has had various parts replaced many times, and looks rather like a cyborg. If it were a human being, it would be a very senior person who had survived many surgeries. Nonetheless, TEPCO was bold enough to describe #1 reactor last year (2010) as "capable of operating for 60 years", and NISA approved this. That is because if a plant stops operation, the electric company sustains a huge loss. The Fukushima Daiichii Plant's reactors were already a danger even before being attacked by the earthquake.

All a Reactor Does Is Boil Water

In what way, then, did this aged Fukushima Daiichii Nuclear Plant, attacked by the Great Tohoku Earthquake, bring about a genpatsu shinsai syndrome? In the introduction I called this a human-made disaster; now I am going to explain why this is so. First I will explain the "heat flow" in a nuclear power plant. Once you understand the plant's mechanism, you will understand how a nuclear power plant disaster will happen and how radioactive substances will leak out.

Nuclear power may sound like state-of-the-art science and something very difficult. However, all a nuclear power plant really does is boil water.

In that sense, it is the same as thermal power generation. The difference is that thermal power plants generate heat by burning petroleum, coal, or natural gas while nuclear power plants generate heat by burning (causing fission in) the element called uranium. Carbon dioxide gas and ash are generated by burning petroleum, coal, and natural gas. Fission in uranium produces radionuclides —that is, death ash. When you compare this system, requiring dizzyingly complex machinery and hordes of experts, with other, simpler ways of boiling water, doesn't it seem a little silly?

Please look at figure 2 (page 26). It illustrates the inside of a nuclear reactor and turbine building. There are two types of reactors: One is the boiling water reactor (BWR), which is mainly used in eastern Japan. The other is the pressurized water reactor (PWR), used mostly in western Japan. In this diagram, the heat flow is illustrated for the case of a BWR, the type used in the Fukushima Daiichii Plant (the basic mechanism to transfer heat is the same for both).

The heart of the reactor where the uranium fuel is located (uranium is put into fuel rods, many of which are bundled together to compose a collective body) is called the reactor core. The reactor core is encased in the reactor pressure vessel, made of thick steel. Outside the pressure vessel is the containment vessel, made of thinner steel. The containment vessel functions to contain radioactivity in a time of emergency.

When the uranium fuel in the reactor undergoes fission, this generates tremendous heat, which is transferred to the water flowing around the fuel, causing it to boil. The steam thus produced is piped to the turbine on the right.

The vapor hits the blades of the turbine, producing kinetic energy. This kinetic energy rotates the generator, and is transformed into electric energy, which is sent away in transmission wires.

How Meltdowns Occur

However, in the process of transforming heat energy into kinetic energy and kinetic energy into electric energy, there is a huge loss. At bottom, the reactor is a clumsy contraption where a third of the heat energy generated by burning (inducing fission in) uranium is turned into electricity and the remaining two thirds is dumped into the ocean as "thermal discharge". Of course, this enormous quantity of thermal discharge raises the temperature of the sea water and greatly effects sea life along the coast. Nevertheless, this contraption is hailed as "the trump card for the prevention of global warming". It's a wonder that there are people who can toss about such falsehoods.

The uranium that burned up in the atomic bomb dropped on Hiroshima amounted to about 800 grams. Currently, there are 54 nuclear reactors on the Japanese Archipelago. One reactor generates around 1 million kilowatts (kW) of electricity on the average. About 2 kilograms of uranium will be burned if a reactor with a daily capacity of 1 million kW is operated for one day. That is to say, one reactor alone each day burns uranium equivalent to three or four Hiroshima bombings.

One of the factors to cause major accidents in nuclear power generation is that the uranium fuel in the reactor core can become overheated and

Figure 2 Heat flow in a BWR

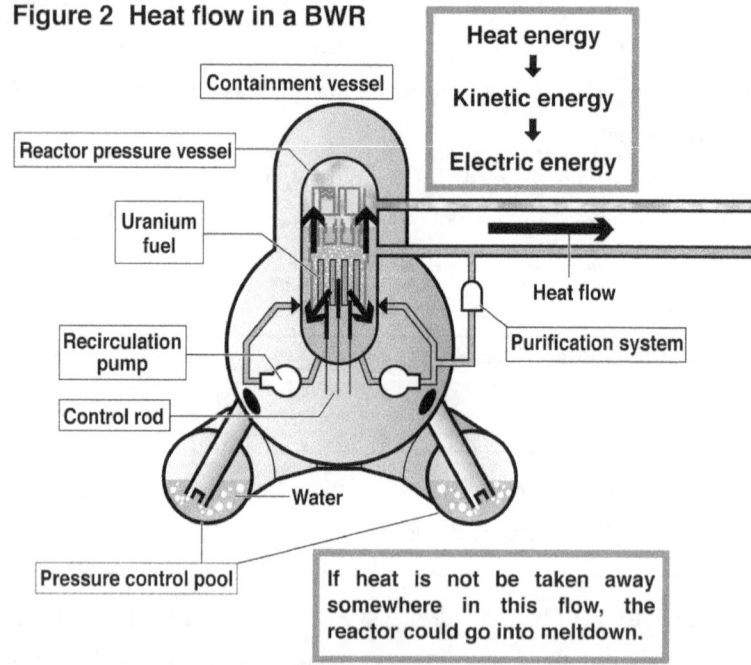

Heat energy
↓
Kinetic energy
↓
Electric energy

Containment vessel

Reactor pressure vessel

Uranium fuel

Heat flow

Recirculation pump

Purification system

Control rod

Water

Pressure control pool

If heat is not be taken away somewhere in this flow, the reactor could go into meltdown.

begin to melt, which is called "core melt" or "meltdown". It is said that reactor cores will begin to melt at 2,800 degrees centigrade. If meltdown were to occur at such a temperature, the heat would easily melt the steel pressure vessel encasing the reactor core as well as the containment vessel surrounding it, so that the fuel would escape outside. Of course, this would include highly toxic radionuclides. Also, when the reactor core melts through the pressure vessel into the containment vessel and touches water there, a hydrogen explosion is very likely. In addition, based among other things on an analysis of America's Three Mile Island Nuclear Power Plant accident , which I will explain later, it has become clear that a core melt may happen at a temperature 600 degrees centigrade.lower than the allegedly necessary 2,800.

The reason the uranium fuel does not become overheated during normal

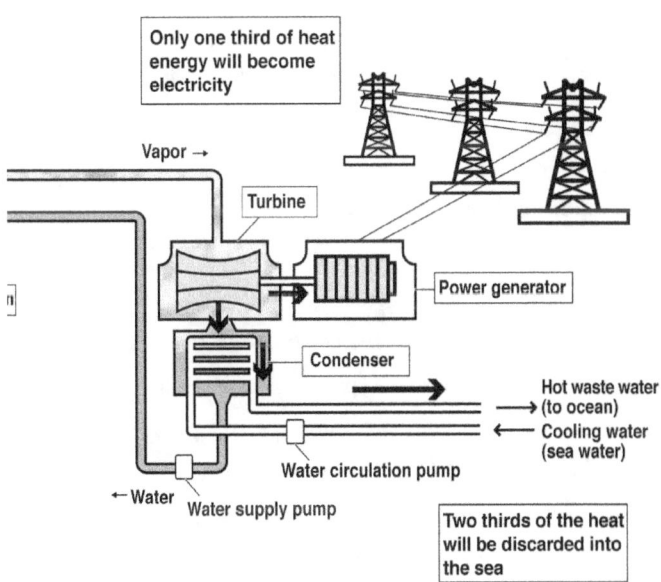

27

operation is that it is cooled by water. As you can see from figure 2, internal pumps operate continuously to circulate the cooling water that carries away the heat. The reason why all nuclear plants in Japan are built on coastal areas is so that sea water can be used as cooling water as the last resort.

Therefore, when an earthquake occurs, even if the reactor itself is built well enough to remain undamaged, if the cooling water fails to circulate, there is the danger of a meltdown. Even if the control rods are inserted immediately to bring the reactor to an automatic emergency shutdown, thus preventing fission from occurring in the fuel, still a large amount of decay heat (this will be explained below) will be emitted.

If the circulation pumps stop working or the pipes are broken and cannot function, the water at the reactor core will soon boil and evaporate unless heat is taken away by some other method. And the uranium fuel rods, which had been submerged in water, become exposed and melt. That is, core melt - meltdown - occurs.

This is exactly what happened at the Fukushima Daiichii Nuclear Power Plant.

The Tsunami Was Not of an "Unimaginable" Height

When the earthquake hit, reactors #1, #2, and #3 at the Fukushima Daiichi Nuclear Power Plant were operating, and reactors #4, #5, and #6 were shut down for a periodic inspection. Having sensed strong shaking, reactor #1, #2, and #3 went into automatic emergency shutdown mode. The situation might have been kept under control if the cooling system had continued to function. At first, after all the reactors shut down, The Emergency Core Cooling System (ECCS) functioned in a normal way and started to cool the system down. Emergency diesel generators began sending electricity to the various pieces of equipment.

However, the situation took a bad turn. Hit by the tsunami, the entire electrical system was destroyed.

Electricity is brought in to nuclear plants on wires carried by steel towers, but with the transmission system broken, all alternating current (electricity from outside) was cut off. The fuel tanks, with the emergency fuel, were washed away. Except for reactor #6, all the emergency power sources for the ECCS were disabled. Some electricity was supplied by emergency batteries, which will last about 8 hours. In order to prepare for such an eventuality, where all outside electricity is cut off and the emergency diesel generators also don't work, still another system is built in. This is the Core Isolation Cooling System (RCIC), which is supposed to generate steam using the decay heat of the reactor core, and use this to operate the turbines which will generate electricity for the pumps. This additional emergency cooling system should have worked, but it stopped after 8 hours because the battery ran out and the direct-current power supply was lost.

Eventually the Fukushima Daiichii Power Plant lost all power sources and its cooling system stopped working entirely. Reactors #1, #2, and #3 went out of control and started to move toward meltdown. A half day after the earthquake (probably by midnight of 12 March) the uranium fuel rods in reactor #1 were already becoming exposed. Even at reactor #4, which was under inspection before the earthquake hit, in the pool where spent uranium fuel rods are submerged in water, the decay heat began to boil the water, and the water level started to fall rapidly. In past nuclear plant disasters –those at Chernobyl and at Three Mile Island in Pennsylvania, USA - only one reactor was involved in each. However, at the Fukushima Daiichii Plant, four reactors went critical at the same time.

An electric power company that lost control of its nuclear power plants because of an electrical failure – it's a pathetic story. But what is unbelievable is that TEPCO put both their normal power source and their emergency power source side by side on the seacoast. It means they did not seriously consider the possibility of a tsunami triggered by an earthquake.

Two days after the tsunami (13 March), Shimizu Masataka, the President of TEPCO, stated at a press conference as follows: "The tsunami was beyond all previous imagination. In the sense that we took all measures that could be thought of for dealing with a tsunami, there was nothing wrong with our preparations." The Fukushima Daiichii Power Plant was hit by Tsunami of a height of 14m, but TEPCO had made preparations for a tsunami of a height of up to 5.5m.

Stop repeating this lie, "beyond all imagination"! The greatest height of the 3/11 tsunami measured so far, that is, as of mid-April, based on the research by Okayasu Akio, Professor at Tokyo Kaiyo University, was at Omoe Peninsula of Miyako-shi, Iwate prefecture: 38.9m. If the investigation continues, a higher record may be found. Looking at the history of plate boundary type earthquakes in Japan, Tsunamis as big as this have often been observed. In the 1896 Meiji-Sanriku Earthquake with an estimated magnitude of 8.5, a tsunami was observed with a height of 38.2m at Ryori, on the Iwate Prefecture coastline; at Yoshihama its height was recorded at 24.4m. In addition, after the 1933 Showa Sanriku Earthquake, a tsunami of 28.7m was observed. After the 1993 Hokkaido Southwest Offshore Earthquake, a tsunami of more than 30m hit Okushiri Island. Can the excuse "beyond all imagination" be allowed to pass? The 3/11 tsunami was slightly higher than any previously recorded, but these records only go back 100 years, and we can easily imagine that much higher tsunamis came in the Edo Era or before.

That a 14m tsunami struck the Fukushima Daiichii Power Plant was not beyond imagination. The melancholy fact is simply that TEPCO did not imagine it. The excuse "beyond imagination" is being used to avoid their being convicted of willful negligence in a court of law. What happened is that the natural disaster revealed their negligence in operating, for 40 years, a nuclear power plant that could easily be destroyed by a tsunami: the natural disaster revealed a human-made disaster.

TEPCO's Fukushima Plant is not the only one that can easily be destroyed by a tsunami. At none of Japan's nuclear power plants, nor at the reprocessing plant at Rokkasho Village, have proper tsunami countermeasures been taken.

Please look at figure 3 (page 32). This is the graph illustrating the tsunami heights that each nuclear power plant has made preparations for. The highest is Hokkaido Electric Power's Kashiwabara Plant at 9.8m. Measures to protect nuclear plants from earthquakes and tsunamis are based on seismic guidelines established by the government. In 2006, these guidelines were revised for the first time in 15 years, and the revised guidelines set down for the first time the measures that should be taken with regard to tsunamis. However we can say about all of Japan's nuclear power plants, that they have no tsunami plans at all. A tsunami on the same scale as the one that hit on 3/11 would destroy any of the plants anywhere in Japan.

Work to Restore Electricity Was Postponed

Having lost its power supply, the Fukushima Daiichii Plant fell into a meltdown crisis. After that the situation, instead of coming together, as if in mockery of the deceptions of the so-called experts (the people who by supporting nuclear power invited this accident) the more time passed, the worse things got. If the root cause of the accident was human-made, so was the worsening of the situation.

Let me explain things in a chronological order. The Fukushima Daiichii Power Plant fell into the worst situation, with radiation squirting out everywhere, because for several days after 3/11 repairs were postponed,

Figure 3 Tsunami preparations of nuclear plants
Estimated heights of largest tsunamis (meter)

Given the heights of past tsunamis, it is clear that all nuclear plants in Japan are vulnerable to destruction.
The Hamaoka Nuclear Plant, which may be hit by the Great Tokai Earthquake, is prepared for a tsunami of only 8m.

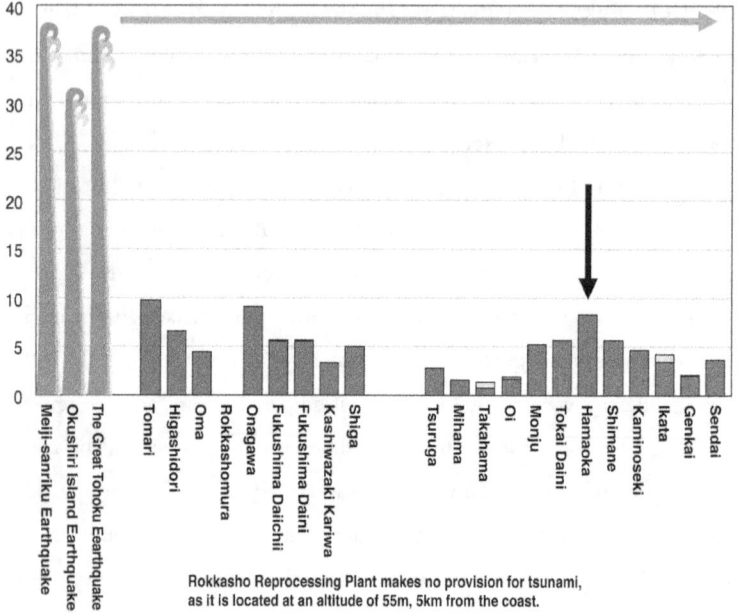

Rokkasho Reprocessing Plant makes no provision for tsunami,
as it is located at an altitude of 55m, 5km from the coast.

and there was a fatal no-policy period. The most frightening fact was their technical incompetence.

After the earthquake on 11 March, I was glued to my television set. When I heard that the cooling system of the Fukushima Daiichii Plant was not working, I watched wide eyed, expecting to see the electrical system repaired, or a truck arrive loaded with diesel generators. Since the cause of the crisis was loss of power, then if only they would only take some emergency measures to fix that . . . .

However, TV "experts" never mentioned the electrical power source, and there was no sign of a truck loaded with power generators on the screen. Why? Does no one understand the cause of the crisis? Executives of

NISA which oversees and supervises TEPCO took an easy-going attitude, saying, "The emergency batteries will last at least 8 hours. The cooling water will not run out immediately and fuel in the reactor core will not be damaged right away; we have about one day of extra time", and never visited the site. I was chilled to think that the lives of our people were in the hands of such as these.

Hardly anybody knows this, but the fact is that this is not the first time for the Fukushima Daiichii Nuclear Power Plant to lose its power source and go critical.

It happened last year. On 17 June 2010 the Fukushima Daiichii Plant #2 reactor lost its electricity and threatened to go into meltdown. It was an accident which should have been big news, but almost no media covered it because they were busy reporting the World Cup Soccer being held in South Africa. It was discovered that all four of the circuits bringing in electricity from outside had been cut off. The frightening thing is that TEPCO never found out the cause of this.

Four days before, on 13 June, a fairly strong earthquake whose epicenter was off the coast of Fukushima Prefecture struck the site of the nuclear station. This could have been the remote cause. At any rate, despite having experienced such a serious accident TEPCO took no countermeasures, NISA did nothing, and less than a year later, we have a catastrophe brought about by the same cause, electricity failure.

If a nuclear power plant loses its power source, the control room, which is the center of the plant's operations, will be blacked out instantly. Even if they know that a reactor is running out of control, there is no way to stop it because even if you press the right buttons nothing happens. This fearsome situation of an across-the-board blackout at nuclear station like this is called "station blackout". In the United States, station blackout simulations have been run, and the Nuclear Regulatory Commission (NRC) has issued strong warnings concerning this danger to all nuclear power plants. In Japan, on the other hand, the station blackout hypothesis itself was abandoned. In 1990 when the Nuclear Safety Commission (NSC) established the safety inspection guidelines for nuclear power plants, the document contained this astounding sentence: "A total and long-term failure of alternating current electrical power need not be considered as quick repair of transmission wires or recovery of the emergency AC system can be expected." In other words, all 54 reactors throughout the nation are operating under the same risk. It is like a car without a brake speeding down a highway.

As far as I have been able to discover, it was 6 days after the earthquake, on 17 March, when TEPCO began in earnest to engage in electricity recovery work (by bringing into the site high-voltage transmission lines from the local power company, Tohoku Electric Power). However, by then, there had been hydrogen explosions in the concrete building where #1 and #3 reactors were housed, another explosion inside #2 reactor's containment vessel, an explosion in the spent fuel pool of #4 reactor, and the situation in general was in shambles – all things the reader well knows.

I will explain the cause of the hydrogen explosions below; here I just want to note that these explosions must surely have given critical damage to the plumbing and electrical wiring systems. In addition the explosions, which blew away the outer buildings, must also have given further damage to the reactors, causing major leakage of radioactive substances. And this made the efforts to restore electrical power even more difficult. TEPCO was (and is) concealing the nature of the accident from the public and, being unable to take proper countermeasures, has been tying a noose around its own neck.

The truth is that after the Great Tohoku Earthquake, the Fukushima Daiichii Plant was not the only one in trouble: at the Tokai Daini Plant in Ibaraki Prefecture one of the three emergency diesel generators was damaged by the sea water, but complete electrical failure was avoided by the other two generators. This was made public by the local paper, Jyoyo Shimbun, on 26 March. Furthermore, on 7 April, almost one month after the accident, the biggest aftershock after 3/11 caused a total blackout in four prefectures: Iwate, Aomori, Yamagata, and Akita. At that time, at Onagawa Nuclear Plant in Miyagi prefecture, where the earthquake was measured at more than *shindo* 6 (the Japanese *shindo* system measures the force of earthquakes at each particular location), two sources of external electric power were cut off, but the remaining circuit managed to continue powering cooling pumps and other facilities. Tohoku Electric's Higashidori Plant in Aomori Prefecture also experienced a power failure; they managed to get it restored before dawn the next morning, in time to avoid a crisis. At the Rokkasho Reprocessing Plant, which is just south of the Higashidori Plant, external power failed and the emergency power source barely succeeded in continuing to cool down the nuclear fuel storage pool and the high-level radioactive liquid waste. Japan was one step short of destruction (this will be explained in chapter 6 in detail).

Why No Help from GE?

You may wonder, Why didn't TEPCO think about electricity recovery earlier? Why did it take so long to get it together? The answer is simple. TEPCO is not an expert in nuclear power.

I use a PC to prepare manuscripts and create graphs. I have a reasonably good command of this electronic equipment. But sometimes the machine suddenly freezes. When the machine is broken, I can't fix it by myself. All I can do is ask the manufacturer to fix it.

The relationship between TEPCO and its nuclear power plants is the same. TEPCO operates nuclear power plants, but they are not the one who built them. Electric power companies are professionals when it comes to wiring, but concerning the complicated structure of nuclear plant design, they are far from being professionals. The same goes for the academics who appear on TV as "nuclear experts". With regard to the details of the reactor design, which differ with each reactor, they are complete amateurs. Academics are not engineers. Thus, if an accident should happen at a nuclear plant, we need to ask the designers at the company that manufactured it for help.

However it has been reported that when Fukushima Daichi went into crisis, TEPCO did not consult with its manufacturer. On the contrary, it is reported that when the manufacturer offered to help, they refused.

The Fukushima Daiichi plant was built by the major US reactor manufacturer, General Electric (GE). Construction began on #1 reactor in 1966. At that time, Japan had not had the experience of building a commercial reactor on its own. Therefore, #1 and #2 were designed by GE, with Toshiba participating in the installation work as a subcontractor following the instructions of GE. (The Actual construction work was conducted by the former Ishikawajima Harima Heavy Industries- currently, IHI.)

It was with the construction of #3 and #4 that Toshiba and Hitachi finally took the lead, but in reality, GE directed the whole project. Reactor #5 is the first one that Toshiba designed on its own. Because of its large size of 1,100,000kW, however, Reactor #6 was once again designed by GE and built by Toshiba. So most of the reactors at the Fukushima Daiichi Plant are of GE manufacture.

Consider what the underside of a reactor actually looks like. Pictures such as Figure 2 should look familiar because they are shown regularly TV and in other media, but this is only a cartoon. The inner structure of an actual reactor, is not as simple as that. In fact beneath a reactor there is a multitude of pipes, electric wires, and valves all tangled together, gauges attached here and there, and control rods dangling down like a forest of spears.

When a place as complicated as this is wrecked by a tsunami or an explosion, surely no one but the one who built it could know what to do. Even more so when the power is down and the place is pitch black, so you have no clear knowledge of how much water is left or how far meltdown has proceeded. I watched the progress of the accident vexed and impatient, thinking that if TEPCO (or the government) refuses the offer from GE, then why don't they start gathering the engineers from Toshiba and Hitachi?

The Advantages and Disadvantages of Pouring On Sea Water

The emergency measure that TEPCO took, before it began work to restore the electricity, was to pour water on the reactors. Since the cooling systems did not work without electricity, their idea was to take away the heat by pouring water on from the outside.

Please take a look at figure 4 (page 38). I mentioned above that even if an emergency automatic shutdown should stop the process of nuclear fission in the uranium fuel, immediately after the shutdown a large amount of decay heat will be emitted. Decay heat is the heat emitted from the radioactive substances generated by nuclear fission of uranium, and it does not soon dissipate. Figure 4 is a graph showing the temperature change.

The graph may be hard to understand since it uses a logarithmic scale of 1 second, 100 seconds, and 1 million seconds. But it shows that even one day after nuclear fission has been halted, more than 15,000 kW of heat will continue to be emitted inside a reactor of 1 million kW.

In case 15,000 kW does not ring a bell, I will calculate this in the form of the Japanese electric heater called kotatsu. [A kotatsu is a small electric heater fixed under a low table, and covered with a blanket.] A kotatsu, with its dial turned up high, uses at most 600w, or 0.6kW, of electricity.

Figure 4
Decay heat after a nuclear reactor shutdown
A reactor with Electric power of 1 million kW = heat output 3.3 million kW

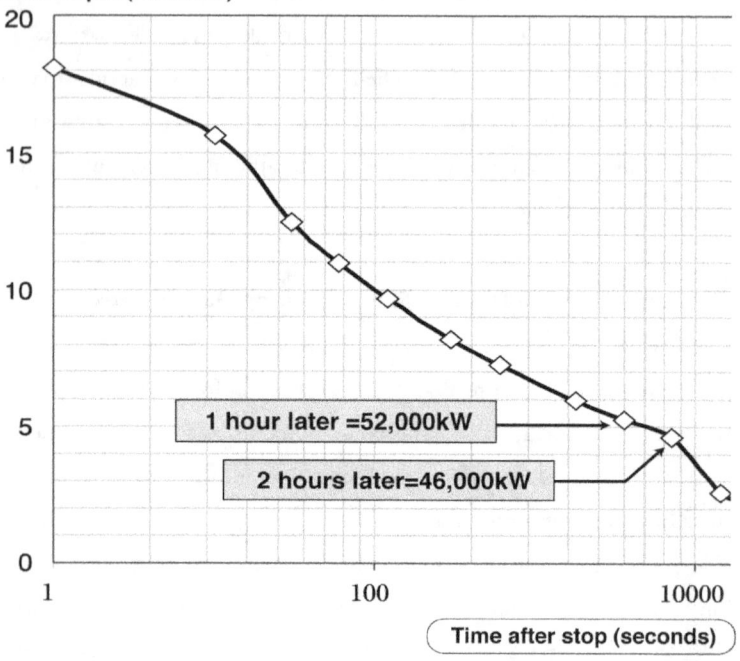

heat output (10000kW)

1 hour later =52,000kW

2 hours later=46,000kW

Time after stop (seconds)

Therefore, 15,000 kW is equivalent to 25,000 kotatsu heaters all turned up all the way. (If you had that much electricity, you could open up a power station.)

Self Defense Force helicopters dangling containers poured water (7.5 tons each) from midair onto the reactors. That is quite literally "pouring water on lava", or to borrow another saying, "pouring water in a flower pot." I was aghast to learn that this was the kind of measure that TEPCO and the government were relying on. Instead of cooling the decay heat, the water would boil and turn to steam instantly.

Sea water was also poured on by fire trucks. This action was proposed to TEPCO right after the earthquake, but it took long time before it

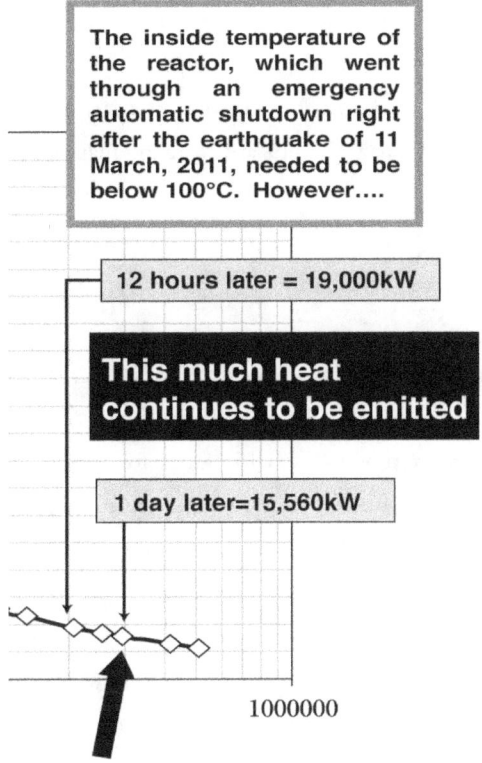

The inside temperature of the reactor, which went through an emergency automatic shutdown right after the earthquake of 11 March, 2011, needed to be below 100°C. However....

12 hours later = 19,000kW

This much heat continues to be emitted

1 day later=15,560kW

1000000

was carried into effect. At the time, a person associated with TEPCO allegedly said "Do you know how much money this nuclear power plant cost? Pouring sea water into it and forcing it to be decommissioned is out of the question." Normally, to decommission a reactor would require 60 billion yen (in case of 1.1 million kW reactor). To build a reactor, much more - several hundreds of billion yen- and long time would be required. Of course, a reactor running out of control will not be put back in order just by dumping sea water over it. It cannot be imagined how many tens of thousands of tons of sea water (later, fresh water) has been dumped in up to now. What it achieved was to destroy all of the reactors, and to flush all the radioactive substances which were inside the reactors out to the outside environment. This is the worst case. I will talk about radioactive substances and exposure in detail in Chapter 3. From the moment they began pouring water on the reactors on a large scale, radionuclides began leaking and today, 27 April, as I write this manuscript, they continue to leak.

The Mark I Reactor Was Flawed Merchandise!

I stated above that the reactors of the Fukushima Daiichii Nuclear Power Plant were made by GE. Reactors #1 through #5 are called Mark I, which it turns out, is a defective product. On 15 March, four days after the earthquake (local time), Dale Bridenbaugh, ex-GE engineer, one of the designers of the Mark I reactor, who retired in protest in 1975 from the company and has been issuing warnings on the danger of the facility, was interviewed on CNN. The Mark I reactor, he said, is not designed to tolerate a large scale accident. In the event of an accident in which the power supply is cut off and the cooling system fails to function, this could result in core meltdown. In a later interview on Fox News he said, "It became general knowledge within GE and within the Nuclear Regulatory Commission in 1975 that all of the Mark I plants were potentially in trouble. They hadn't been designed to withstand the accident sequences they were supposed to have been designed to. . . [I said to my boss] we've got to do something, we've got to commit to shutting these plants down if we find the analyses we are getting right now are correct. And my boss said, we can't do that. If we shut down all of these plants – there were 16 in operation at the time – if we shut those down now, this could mean the end of GE's nuclear program."

So in order to save GE's nuclear power plant program, a plant known (by GE management, and presumably by TEPCO as well) to be flawed was built at Fukushima.

As a result, the capacity of the containment vessels and suppression chambers of Mark I reactors is a half to a fifth that of the Mark III. The Mark I containment vessel is designed to tolerate pressures of up to about 4 atmospheres (atm).

However following the earthquake, because its cooling system failed, inside the Mark I #1 reactor at Fukushima Daiichii the water level fell. That is, the water was boiled away by the decay heat, and gas (steam) started to fill in the area, which was sent to the containment vessel, where the pressure rose to as high as 8 atm, close to bursting.

What did TEPCO then? Please look at figure 5. As they had to decrease pressure inside containment vessel, they opened the relief valve and released gas outside the building. It goes without saying that the gas

Figure 5 Method to lower pressure of containment vessel in times of emergency

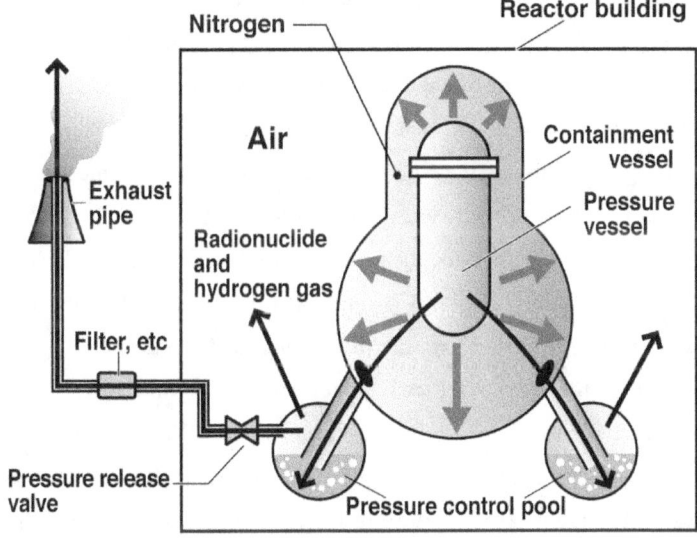

Since containment vessel will be ruptured if inside pressure gets higher, high pressure gas is released outside along with radionuclides by opening the valve.

contained an abundance of radioactive substances. TEPCO intentionally released radioactive substances from the containment vessel.

Shortly after that was done, at 3:36 pm of 12 March, with a great crashing sound a hydrogen explosion occurred in the building containing #1 reactor. All the concrete of the upper building was blown away and the containment vessel was exposed.

You know by now what this means.

The Earthquake Destroyed the Water Pipes with One Blow

In the last section of this chapter, I am going to add a frightening story that television and the newspapers have hardly covered.

There is a person who has been trying to solve the puzzles surrounding the way the accident unfolded, including the question, By what route did the hydrogen enter and fill the operation floor of the reactor building, causing the massive explosion? This is Tanaka Mitsuhiko, ex-chief designer of #4 reactor of the Fukushima Daiichii Power Plant. Tanaka is a science writer and has written exposes on the danger of nuclear power plants. While analyzing the sequence of events leading up to the accident, he found a very important fact in documents from the office of the Prime Minister. The pressure inside the pressure vessel of #1 reactor, which had been at its operating level of 70 atm, when measured at 2:45 am on 12 March (that is, just half a day after the earthquake) had plunged to 9 atm.

This means that among the many pipes connected to the reactor, some must have been damaged and pressure must have been hissing out. A likely possibility is that, for example, the huge recirculation pump that hangs on a side of boiling water reactors was shaken by the first direct shock of the earthquake and its plumbing was damaged. Or, it could have been some other parts. The damaged part is unknown since now nobody can look inside. But there is no doubt about the fact that the pressure escaped.

That is, somewhere in the pipes there was an opening, through which the steam in the reactor blasted into the containment vessel outside. Consequently, the fuel rods raised their heads up above the water. As the reactor went into meltdown, zircalloy (an alloy of zirconium) in the pipes encasing the fuel rods reacted with the water and began to oxidize, and the oxidization, by taking oxygen out of the water, released hydrogen gas. This hydrogen flowed out through the damaged pipes and, being light, accumulated in the upper portion of the containment vessel. This caused the interior pressure of the containment vessel to rise to a dangerous 8 atm, enough to lift a lid and allow it to escape and fill the concrete building called the operation floor. Then, reacting with the oxygen in the air there, the hydrogen exploded. This is a most plausible case. Tanaka is asking, why did the pressure in the reactor drop, and why did the pressure in the containment vessel rise? If there is some other scenario to explain this, he wants the government or TEPCO to say what it is. Because if there was a hole like that, then the cooling operation would not have worked even if electricity had been restored and water had begun to circulate. And that would be the beginning of the process by which radiation contamination is spreading throughout eastern Japan.

The main reason this is being hidden is that if it is revealed, that could mean that the seismic safety back-check that NISA and NSC are presently carrying out with Japan's 54 reactors has no meaning. The force of the earthquake was finally made public by TEPCO in its website on 1 April. If they were not lying, from #1 reactor to #6 reactor, the ground acceleration was east-west at no more than 319 to 550 Gals (Gal – or Galileo - is a unit to indicate the actual force of a quake on the ground). If the plumbing was damaged by as small scale a quake as this then, for example, the Great Tokai Earthquake, which is threatening the Hamaoka Nuclear Power Plant in Shizuoka Prefecture, would cause an immediate and fatal catastrophe. (I will explain about this emergent crisis in the Chapter 5 in detail.)

As other nuclear experts supported Tanaka's speculation, TEPCO began urgently to deny it. They offered a different explanation: the pressure dropped because the valve for releasing steam from the pressure vessel got stuck in the open position. So the steam escaped because a valve was open – that is intended as an excuse. But with this explanation they are admitting that the reactor suffered a loss of coolant, an accident that should not happen under any circumstances. If that is what did happen, that in itself would be an extremely serious accident, one that completely refutes all the volumes that have been written about the safety of nuclear reactors. But TEPCO was unable to provide any evidence that that is what happened, and in the end admitted that they could not deny the possibility that damage to plumbing and other systems could have been the cause.

Chapter 2: What TEPCO and the Media are Hiding

When did the radiation start to leak from Fukushima Daiichi Nuclear Power Plant and how far has it spread?

About this our greatest interest, the Japanese Government announced nothing for several days (at least until 17 March) after the earthquake occurred on 11 March, 2011. They just described the facts saying they "....happened" in the past tense, explaining neither the cause of the explosion nor its mechanism. They never talked about expectations of danger. What is "the worst scenario" that this nuclear accident might lead to? -- A reporter asked this of Edano Yukio, Chief Cabinet Secretary, at a news conference, but whether because Edano couldn't answer or didn't want to, he said nothing clearly.

Like the government, most of the media, rather than informing us of the dangers posed by this accident, kept claiming groundlessly that there was none. The stock phrase used by both Edano and the TV "experts" - "no immediate effects to health" – means what? That radiation effects start one year or 10 years later? If the radiation were to have "immediate effects", that would be a horror. People would be dropping dead in the streets of acute radiation sickness as they did after the atomic bombings of Hiroshima and Nagasaki. But that is not what we are talking about. The question is whether radiation taken into children's bodies is really safe.

The "Experts" Who Were Unable To Foresee a Hydrogen Explosion

This is the first time a genpatsu shinsai syndrome occurred in Japan – or in the world. With the situation changing by the minute, people surely would want to know more about what was likely to happen next, and not just what had already happened at the Fukushima Daiichi Plant.

But was there any media that was able to foresee what was coming?

On the afternoon of March 12, the day after the earthquake, a hydrogen explosion occurred in #1 reactor at Fukushima Daiichi Plant and the reactor building suffered severe damage. As I mentioned in the last part of Chapter 1, this explosion should have been avoided at any cost, because a large amount of radioactive substances was carried away by the blast wind and spread into the atmosphere.

It is because neither the government, nor the Nuclear and Industrial Safety Agency (NISA), nor the Nuclear Safety Commission (NSC), nor TEPCO was able to predict the explosion that they failed to avoided it. I was watching this process on TV and thought of the 1979 Three Mile Island accident. Remembering the effect of decay heat, I was on edge, thinking "There's going to be an explosion!" I was praying those "experts" on TV would raise the alarm and that the danger could be avoided. But no one said a thing.

The cause of the hydrogen explosion is simple. In #1 reactor the cooling system stopped working just after the quake, and after a full day with no cooling most of the uranium fuel rods in the core must have melted down. Uranium fuel is processed into the shape of pellets and encased in zircalloy (an alloy of zirconium) pipes.

As I explained in Chapter 1, if the core starts to melt, the zirconium, making contact with water, reacts, and oxidization begins. If the metal oxidizes in the water (removing the O from the H_2O), more and more hydrogen is generated. If enough hydrogen is produced, explosion occurs.

Just when I was getting more and more irritated watching the TV where no one was mentioning this, the phone rang. It was a newspaper reporter asking for my comments.

I said, "Go immediately to the NISA press conference and ask them about the danger of a hydrogen explosion."

Right after that, the explosion occurred.

After the explosion those "experts" were on TV again, giving plausible explanations of what had caused it. This is what we call showing up for the fair a day late. Real experts would be ashamed of not being able to explain the danger of the explosion before it happened.

At the press conference after the explosion, Chief Cabinet Secretary Edano said, "Even though #1 reactor building is damaged, the containment vessel is undamaged. On the contrary, the outside monitors show that the [radiation] dose rate is declining, so the cooling of the reactor is proceeding."

Both NISA and NSC insisted, "For this accident to reach the Chernobyl level is out of the question."

Most of the media believed this and the university professors encouraged optimism. It makes no logical sense to say, as Edano did, that the safety of the containment vessel could be determined by monitoring the radiation dose rate. All he did was repeat the lecture given him by TEPCO. I was persuaded that TEPCO, the government spokesmen and the TV commentators are a pack of amateurs. I had a bad feeling that something even worse was going to happen.

It was on 14 March, two days after the explosion at #1 reactor, that the hydrogen explosion occurred at #3 reactor. Though it was the second explosion I had not seen any "experts" on TV who predicted it or warned about it.

At Three Mile Island, an Explosion Was Avoided

In case of the March, 1979 Three Mile Island Nuclear Power Plant accident, the Emergency Core Cooling System (ECCS) functioned well after the cooling pump stopped. But the operator, confused by the water level, stopped the system and a large amount of water was lost, causing a very serious situation. With two-thirds of the uranium fuel rods exposed it was like putting a kettle with no water over a fire, and meltdown started. However, in this first major accident in the history of nuclear power plants, there was no explosion..

As at Fukushima, there was danger of a hydrogen explosion at Three Mile Island but the brains at the Nuclear Regulatory Commission (NRC) prevented it.

The explosion threshold for hydrogen is 4.2%. That means if hydrogen reaches this concentration in a closed chamber within which there is a gas that contains oxygen, it explodes. The NRC predicted the danger of a hydrogen explosion in the reactor building in the Three Mile Island Nuclear Power Plant and calculated the level of concentration, keeping it at just 1% short of the threshold, thus avoiding the explosion. The NRC is an independent regulatory agency of the federal government which approves and licenses nuclear power facilities in U.S. In Japan there is no specialized nuclear power agency with that kind of know-how. The various agencies are just groups of people with no idea what to do in case of trouble.

Returning to the story in Japan, I will explain the process of the hydrogen explosion at #3 reactor, which occurred after the one at #1 reactor. The process was as follows.

It began on 13 March, the day after the #1 reactor explosion. Just after 7 AM, the feed-water stopped and #3 reactor lost its cooling function completely. With three-quarters of the fuel exposed meltdown began, and the hydrogen explosion occurred through exactly the same mechanism as that of #1 reactor. The time was 11:01 AM on March 14. It took about a day from the time the cooling function stopped until the explosion. That was about the same amount of time as it took #1 reactor. But after the explosion the situation got worse than at #1 reactor. The building was blown off together with its steel frame, and 4 TEPCO employees and 3 workers from partner companies were injured.

Seeing #3 reactor damaged so severely, I had another chilling premonition. Although it was hardly reported in the media at the time, this #3 reactor was a "special" reactor.

Afraid To Mention Pluthermal?

Reactor #3 was the only one in which "pluthermal" power-generation was taking place at Fukushima Daiichi Plant. It was August of last year (2010) that I went to Fukushima Pref. to give a lecture against pluthermal power generation. Pluthermal is a way to generate electricity using nuclear fission of plutonium fuel in a nuclear reactor built for uranium. The term is a Japanese-English neologism, combining "plutonium" and "thermal neutron reactor (light-water reactor)".

I will explain plainly.

The element uranium consists of fissionable U-235 and non-fissionable U-238. U-235 is very scarce, accounting for only 0.7% of natural uranium ore. The rest, accounting more than 99%, is U-238, that is, non-fissionable uranium.

However, U-238 has a specific property. When uranium fuel is put into a reactor and U-235 is burnt there, the U-238 around it absorbs neutrons and turns into fissionable plutonium-239. The U-238, which accounts for more than 99% of the volume, while not fissionable, turns into plutonium. Thus from the spent fuel left after the uranium has been "burned", plutonium can be extracted.

Pluthermal uses this plutonium-uranium compound oxide, processed into fuel. Called "MOX (mixed oxide) fuel", it is used in reactors where it generates thermal energy by fission which in turn is transformed into electric energy. So pluthermal recycles spent nuclear fuel, and transforms non-fissionable U-238 into fissionable plutonium, leading some experts to claim that a "nuclear fuel cycle" has been effectively achieved. In fact, however, the cycle hardly goes round at all.

The problem is that this plutonium is highly poisonous; it is said to be "one of the most toxic substances that human beings have ever encountered". Plutonium is extracted from spent nuclear fuel; it does not exist in nature. It is a man-made element, originally created to make an atomic bomb.

Having this character, plutonium-based MOX fuel emits a larger amount of radioactivity than ordinary uranium fuel. If we should breathe even a few particles of this highly poisonous plutonium into our body, they will stay there for a long time and we will be exposed to strong and continuous radiation (alpha rays). (This is called "internal exposure". I will explain it in detail in Chapter 3.)

Obviously, plutonium requires extremely careful handling, however the Japanese government has declared that it is going to use it to realize a nuclear fuel cycle (in fact, impossible), and in 2005 adopted the Framework for Nuclear Energy Policy, which projects the start of pluthermal operation in 16 to 18 reactors by 2015. In December 2009, the first commercial pluthermal operation started at Kyushu Electric Power Company's Ikata Nuclear Power Plant (Ikata-cho, Ehime Pref.). And then on 26 October 2010 (the year before the earthquake), commercial pluthermal operation started in No.3 reactor at Fukushima Daiichi Nuclear Power Plant. Number 3 was an aging reactor, used for 34 years, but it quickly gained public attention because of the pluthermal operation. What did the Government officials think when they saw a hydrogen explosion occur at #3 reactor less than six months after the pluthermal operation started? The Democratic Party of Japan (DPJ) Government has been even more active at promoting nuclear power than the Liberal Democratic Party (LDP) Government that preceded it. On 31 October 2010, just after pluthermal operation had started at No.3 reactor, Prime Minister Kan Naoto visited Vietnam with Maehara Seiji, then Foreign Minister, and met with Prime Minister Nguen Tan Dung in Hanoi. They agreed that Japan would undertake the construction of two nuclear power plants and Kan was all smiles. Of course both pluthermal operations and the power plants in Vietnam must be scrapped.

After the explosion at #3 reactor, I worried whether the extremely dangerous radioactive substances had escaped outside. It turned out that they had. TEPCO has announced that, on 21 and 22 March, in soil collected on the plant site, they detected plutonium.

Was the Earthquake Really Magnitude 9?

Aside from the way the genpatsu shinsai syndrome was reported, there was another thing I could not understand about the media report of the Great Tohoku Earthquake. It is the expression "once-in-a-thousand-years earthquake" used by every TV station including NHK (Japan Broadcasting Corporation). The expression is based on the figure "magnitude 9" announced by Japan Meteorological Agency (JMA). But was it really that big an earthquake?

When an earthquake occurs, it is recorded by seismographs, and the amount of energy generated at the epicenter is called the magnitude. We can also call it the "energy of the seismic wave released from the epicenter". If magnitude figure is 0.2 larger, that means the energy is about twice as great, so if the magnitude is greater by a factor of 1.0, that means the energy is 32 times greater (2 to the fifth power: 2x2x2x2).

Accordingly it can be said a that if the Northeastern Sanriku Offshore Earthquake had a magnitude of 9, that would mean its energy was 45 times that of the magnitude 7.9 Kanto Earthquake which brought about the Great Kanto Earthquake Disaster on 1 September, 1923. I can hardly believe this. The epicenter of the Kanto Earthquake was 80 km off north-west Sagami Bay, Kanagawa Pref., so it did not hit Tokyo directly. In the Southern Hyogo Prefectural Earthquake which brought about the Great Hanshin Earthquake Disaster, the magnitude was recorded at 7.3. That would calculate out to mean that the energy of the Great Tohoku Earthquake was 355 times that of the Southern Hyogo Prefectural Earthquake.

When I talked with news people who reported on the disaster, they all said with one voice, the earthquake damage was not so great, but the tsunami was appalling. I myself saw the enormous damage wreaked by the tsunami on TV and was terrified. However, the damage from the quake itself is not clearly known. One reporter said that this time he did not see anything like what he saw during the Southern Hyogo Prefectural Earthquake, when the Hanshin Expressway was toppled off its foundations.

So I examined the data.

"Gal" is a unit for measuring the strength of a quake precisely. It is an acceleration unit named after Galileo. One Gal indicates the amount of force which, when it acts on an object 1 gram in weight, will cause that object to accelerate at a rate of 1 centimeter per second. In time of earthquake this rate of acceleration indicates the momentary force that acts on people and buildings.

In the Great Tohoku Earthquake the strongest tremor, as far as I know, was the *shindo* 7 recorded at Tsukidate Kurihara City, Miyagi Pref. [again, the *shindo* is a unit used in Japan to measure the force of an earthquake at each specific site, not at the epicenter]. The ground acceleration figure there was 2933 Gal. A big quake it is, but in the Iwate Miyagi Inland Earthquake (magnitude 7.2) in 2008, they recorded vertical acceleration of 3866 Gal and three-component synthesis maximum acceleration of 4012 Gal (the highest ever recorded) in Ichinoseki City, Iwate Pref. This would mean that while its magnitude was less than that of the Great Tohoku Earthquake, its force on the ground was greater.

Was the 3/11 earthquake really the most powerful earthquake in history? Watching the process, it seemed suspicious. At first JMA announced the magnitude of the earthquake provisionally as 8.4. Then this was corrected to 8.8, and finally "upgraded" to 9.0.

Escaping Responsibility by Managing the News

The seismologist and geologist Shimamura Hideki (former Director of the National Institute of Polar Research) said, "This never-before-heard-of figure of magnitude 9 is was produced by JMA's arbitrary altering of the magnitude scale.

In fact a number of different scales have been used to calculate magnitude. In Japan in the past the "JMA magnitude (Mj)"scale has been used. If we enter data of the Great Tohoku Earthquake into the Mj formula (omitted here as it is complicated), its magnitude would be 8.3 or 8.4 at most according to Shimamura. He continues, "It is because JMA recalculated the data using the 'moment magnitude (Mw)' scale, which has been used only by scientists, that the magnitude could be raised to 9.0."

Note that in this book the magnitude of all the past earthquakes is shown by the customary Mj scale. If the scale is changed abruptly, it becomes impossible to compare with past earthquakes. Why did JMA change the scale without explanation? Here I can feel political intervention distorting a scientific truth. If the magnitude figure remained at the original 8.4, that would mean that the earthquake fell within the range of a "predictable" disaster. And if that were so, not only TEPCO, but also the Japanese government and the "experts" who have promoted nuclear power, would be held responsible.

As I mentioned in the introduction, The Chubu Electric Power Company (CHUDEN) has announced that its Hamaoka Nuclear Power Plant is built to withstand an earthquake of a magnitude of magnitude 8.4. Advocates of nuclear power needed to raise the magnitude to 9.0, otherwise they would be in a pinch. By announcing this unprecedented figure, they want to make people think the earthquake was "a once-in-a-thousand-years earthquake" and "beyond expectation" thus escaping responsibility.

Especially TEPCO has tried to avoid criticism by raising the magnitude to 9.0. Even now their sales are decreasing sharply because of planned outages and electricity conservation measures by consumers. If they need to pay huge damage compensation for the Fukushima Daiichi Nuclear Plant accident, it will be hard to keep the company afloat. They want to keep compensation costs as low as possible; that is their aim as a company. Conveniently for them there is the following law:

Act on Compensation for Nuclear Damage

Part 2 Liability for Nuclear Damage

(liability without fault, channeling of liability etc.)

Section3 When a nuclear reactor is operating, and when as a result of such operation nuclear damage is caused, the operator of that reactor on that occasion shall be liable for the damage. However in the case where the damage is caused by a grave natural disaster of an exceptional character or by an insurrection, this shall not apply.

This is the Nuclear Power Plant Damage Compensation Law. If one translates what it says into ordinary language, it reads, "If a nuclear power accident should occur as a result of an unpredictable disaster, the responsibility for damage compensation will be placed not only on the company but also on the government and taxpayers".

Consequently for TEPCO, if the magnitude is raised from 8.4 to 9.0, their responsibility for damage compensation can be reduced. On 30 March, TEPCO chairman Katsumata Tsunehisa showed up at a news conference for the first time after the meltdown. He said, "We would like to offer as much apology and compensation as possible". But as to who would be covered and how much it would be, he said "We will discuss that with the government".

Now there a danger that an outrageous demand will be made on the victims of this disaster, the Japanese people. The principal victims are the people who have been forced to evacuate from their homes around the plant and farmers of products whose shipments have been suspended. The damage caused by leaking radiation has spread to a wider area and into many regions, disrupting our ordinary lives. However, I foresee an unforgivable ending in which the damage brought upon the people by the negligence of TEPCO will be paid for by the people's tax money.

Chapter 3: Long Battle with Radiation

The next topic will be that of exposure to radiation, the most important and life-threatening of all. When we discuss the question of safety with regard to nuclear power plants, we inevitably come to the question, what level of radiation exposure is safe for human beings and what level is dangerous?"

Ever since the Fukushima Daiichi Nuclear Power Plant accident, the media including TV have rushed to report that radiation exposure can be given a numerical value. Are those numerical values reliable? Who determined them? It is not easy for ordinary people to understand the meaning of these figures.

Both before and after the recent accident, the media have been overwhelmingly deceptive, because the discussion always focuses on these numerical values rather than on the most fundamental of fundamental questions, what is the danger of radiation? One of the major reasons for this is the fact that a generation of young people who hardly know the meaning of "radioactive fallout" is at the center of today's media. Also, those "specialists" who are promoters of nuclear power, though their titles may include the names of famous medical institutions, never allow words referring to radiation risk to pass their lips. At the same time, they cannot present any valid numbers that would support the case that there is no danger. These are the people who have earned their living off the notion of "efficient" use of radiation.

Let me make it clear: I do not believe in these numerical values for measuring radiation danger.

Where Does Radiation Come From?

Let's talk about some of the basic facts concerning radioactivity. Radioactivity means "ability to emit radiation" and sometimes stands for the substances that have that ability: radioactive materials or radionuclides. The most distinctive feature of radioactivity is that it is extremely dangerous to the human body even though it cannot be detected with the five senses. When we are shot by a gun, we feel pain. When we inhale cedar pollen we get a runny nose. However, when we are exposed to radiation (this is called irradiation) we feel nothing. It may happen that such exposure may lead us, feeling nothing, to our death.

In the first chapter, I explained the mechanism by which uranium fuel in a reactor undergoes fission, produces intense heat and generates electricity. The "nuclear fission of uranium fuels" means, scientifically, that the uranium atom, bombarded by neutron radiation, is split into two or three parts (the uranium atom is big while the neutron, being one of the particles composing the uranium atom's atomic nucleus, is small).

When the uranium atom is split, smaller atoms will be generated. However, these atoms are irregular in shape, and unstable. They become radioactive substances such as iodine 131, cesium 127 and strontium 90 and start emitting radioactive rays called alpha rays, beta rays, and gamma rays. Also, as I explained in Chapter 2, the nuclear fission of uranium generates a radionuclide called plutonium which emits alpha rays.

Each kind of radioactive ray emitted from radioactive materials has different characteristics. Their penetrating power is different.

Alpha rays have lower penetrating power and their penetration can be blocked with a sheet of paper. Beta rays can penetrate paper, but can be stopped by metal.

Gamma rays are the same as x-rays used at hospitals (roentgen) and can go through normal metals. They can be stopped by concrete or lead walls. However, gamma rays generated from, for example, cesium 137 and cobalt 60 have a higher penetration power. Especially cobalt, which is used in radiation therapy for cancer, penetrates deep inside the human body and thus is very dangerous. That is why pregnant women need to avoid having x-rays taken.

Neutron rays are even more dangerous. Being electronically neutral, they are not affected by the electronic resistance of substances. Therefore, they penetrate most anything. There was a time when scientists, making ill use of this feature, were trying to develop a neutron bomb that would destroy living creatures without destroying any buildings.

Tricks to Make It Sound Safe

Many readers will have heard the term "half-life" on TV or elsewhere. TV "experts" tell us for example that since the half-life of iodine 131 is around eight days, the risk of exposure to radiation will decrease over time. But they leave out the most important point, which is that once it has escaped, iodine 131 will never go away.

Half-life is the time required for the radioactivity of a radionuclide to decrease by half. When the first half-life period passes, the original radioactivity of a substance is reduced to half, then to a quarter, then an eighth, a sixteenth…etc. However, no matter how many times this calculation is repeated, it never reaches zero. As I will explain later in detail, if radioactive particles are floating in the atmosphere, and if even a few of these should enter a human body and cause internal irradiation, the body may be affected even after many years have passed by.

There are radioactive substances whose half- life is very long. For example, plutonium 239, which was mentioned above and which was used in the atomic bomb dropped on Nagasaki, has a half-life of 24,000 years. Even after 140,000 years, when it is not clear if we human beings will still be around, plutonium will still have one sixty-fourth of its radiation.

Then, are radioactive materials with a shorter half-life safer? Not necessarily. Radioactive substances with a shorter half-life generate more energy in a short period of time. For example, the half-life of plutonium 238 is 87.7 years, which is shorter than that of plutonium 239; however, its radioactive energy is about 270 times greater. The relative danger or safety of a radioactive substance cannot be determined by its half-life.

As mentioned in Chapter 2, on March 15th, the day after #3 reactor at the Fukushima Daiichi Nuclear Power Plant underwent a hydrogen explosion, Chief Cabinet Secretary, Edano declared that 400 mSv per hour of radioactivity had been measured around the periphery of #3 reactor, and admitted for the first time that this was a "life threatening figure". However, the meaning of this figure is hard for the general public to interpret. How dangerous would this be? Our government has no idea because they have no medical understanding of the human body.

The Sievert (Sv) is a numerical value given to the effect that radioactivity has on the human body: 1 Sievert = 1,000 millisieverts (mSv) = 1,000,000 micro Sieverts (symbolized with the Greek letter mu + Sv). The International Commission of Radiological Protection (ICRP) stipulates that the maximum annual exposure limit for the general public should be 1mSv. The figure given by Mr.Edano is 400 times that exposure, each hour.

Chief Cabinet Secretary Edano's calculation by the hour and the ICRP's calculation by year are not comparable. The 400 mSv per hour figure must be recalculated to show what it means per year: 365 days x 24 hours = 8,670. Multiplying this by the 400 mSv per hour, we find an annual risk of about 3.5 million times the permissible amount, which is outrageously dangerous.   This is not a figure that can be described casually in the phrase "will affect the human body".

Unlike the rule for the general public, the upper limit of annual exposure for workers at facilities including nuclear power plants is designated as "up to 50mSv" by the Ionizing Radiation Obstacle Prevention Rule. However, Defense Minister, Kitazawa Toshimi on March 15 approved an upper limit of 100 mSv for TEPCO workers and Self Defense Forces (SFD) who are engaged in emergency assignments. He further announced that, as long as the Atomic Energy Emergency Declaration is in effect, the limit would be further increased to 250mSv. This amounts to saying that workers sent in to these accidents sites are on a mission for which they must be prepared to die.

The Fukushima Daiichi accident teaches us that what is overlooked is sacrifice of the workers who, in the urgency of the situation, and facing real danger, willingly go on working at the accident sites. It also teaches us that, in contrast to them, the government men and TEPCO executives and TV "experts" who talk importantly about these things in Tokyo are not to be believed.

Internal Exposure, Which Cannot Be Measured by Numerical Values

"No immediate health effect" is almost always followed by the set phrase, "we are irradiated when we have a stomach x-ray or fly in an airplane and are exposed to cosmic radiation".

I want to make clear that to bring up such examples as this in order to argue the safety of minor levels of radiation exposure is close to fraud. In the first place, exposure to X-rays and cosmic radiation are temporary situations. It makes no sense to calculate such figures into annual exposure rates and compare them with the upper limit discussed earlier. In the second place, exposure to radiation emitted from the nuclear

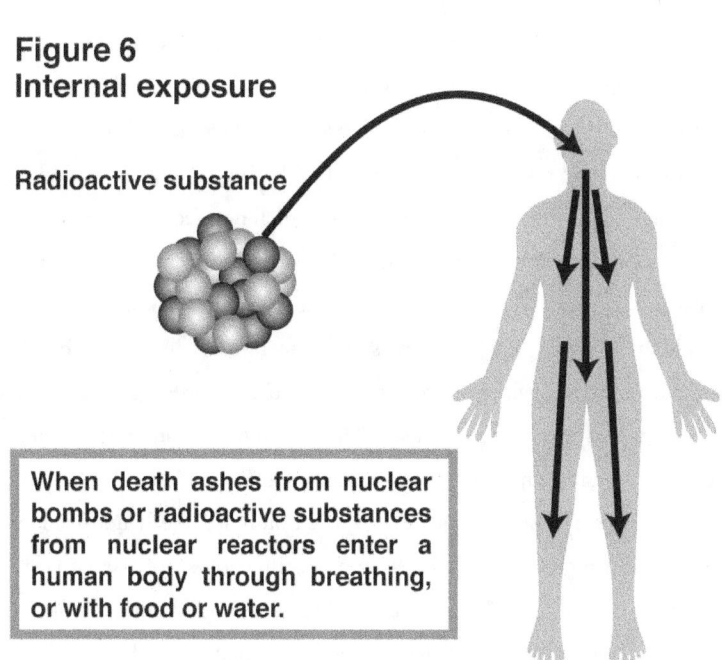

Figure 6
Internal exposure

Radioactive substance

When death ashes from nuclear bombs or radioactive substances from nuclear reactors enter a human body through breathing, or with food or water.

plant accident will be added on to past exposures, so it is misleading to "compare" the figures.

Another off-the-mark safety argument is the story that goes "the strength of radioactivity is inversely proportional to the square of the distance, so locations distant from the Fukushima Daiichi Nuclear Power Plants are safe." Radioactive materials (as opposed to radioactive rays) are carried by the wind. In particular, iodine and cesium are easy to spread. If they are carried to places far from the Fukushima Daiichi Plants, they will release radioactivity there. In other words, even at places distant from the accident site, if radioactive substances come near, risk of exposure increases.

In the Chernobyl accident #4 reactor, then in operation, went out of control and ended up exploding on a large scale. The Soviet Union Government authority evacuated people who lived within a 30 km range from the accident, making a no-man's land. Afterwards, however, areas more than 200km away from the accident were found to be highly contaminated by radioactivity (hot spots). It is believed that these hot spots were created by radioactive substances which are released into the atmosphere by the explosion traveled with clouds and descended with the rain. 200 km is approximately the distance from the Fukushima Daiichi Nuclear Power Plant to the Tokyo area. In government statements and media reports relating to exposure after the Fukushima Daiichi accident, the biggest problem is that the figures discussed there are mostly numerical values given to radiation doses. These reports are always accompanied with words such as, "This is a miniscule amount, so there is no worry", and "There will be no immediate effect on health," which most people believe to be true. However, those claims are utterly nonscientific. Measuring instruments such as monitoring posts are used to gauge radiation amount. Such instruments are placed around nuclear plants and other facilities to measure radiation coming to them from the outside.

However, there are two types of exposure. What can be measured using monitoring posts and counters is the radiation a human body would be exposed to when the radioactive substances are outside it; that is, they measure the danger of external exposure. On the other hand, what is more terrifying in nuclear accidents is internal exposure. This is occurs when radioactive substances are taken into the human body for some reason, and its numerical value cannot be measured by a monitoring post. Machines do not eat, nor do they inhale, radioactive substances. The government and the media's "no immediate effect" does not take "internal exposure" into account.

Japan's Radiologist Community: The World's Most Obtuse

Internal exposure takes place when radioactive substances enter the human body via the mouth along with food or water or via the mouth

Figure 7
Difference between external and internal exposure

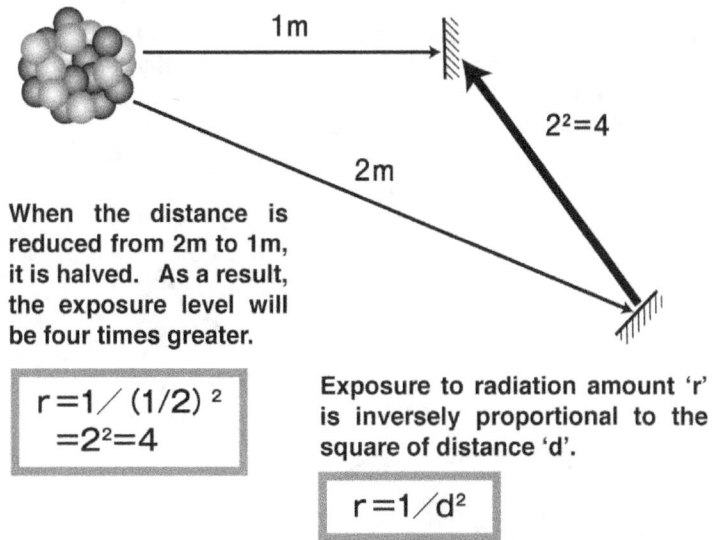

1m

$2^2=4$

2m

When the distance is reduced from 2m to 1m, it is halved. As a result, the exposure level will be four times greater.

$$r = 1 / (1/2)^2$$
$$= 2^2 = 4$$

Exposure to radiation amount 'r' is inversely proportional to the square of distance 'd'.

$$r = 1 / d^2$$

or nose through breathing. In that case, the person will continue to be irradiated from inside the body by those substances.

The media has been reporting that "the strength of radioactivity is inversely proportional to the square of the distance, so places far from the nuclear site have no problem. " However, what happens when we apply this inverse-proportional logic to the internal exposure case? Here the human being is irradiated from the closest place possible (inside the body). And the closer the distance from the radiation source, the greater the exposure will be (figure 7).

Suppose the radioactive substance enters the lung and sticks to cell tissues there. If we assume that the substance is separated from the tissue by a distance of 1 micron, compared to the case when the substance is located outside the body and one meter away, what will be the difference in exposure? One meter is 1,000 mm, 1 mm is 1,000 microns. Thus, the

Figure 8
Internal exposure of Plutonium

Alpha rays emitted from plutonium has mass and is positively charged. Therefore, they normally travels only a short distance (less than 1mm) in water. Even a sheet of paper can stop alpha rays.

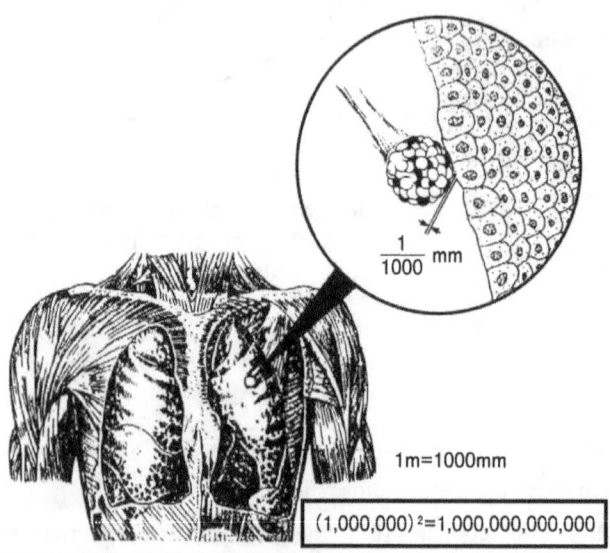

$\frac{1}{1000}$ mm

1m=1000mm

$(1,000,000)^2 = 1,000,000,000,000$

However, if the plutonium is inhaled into a lung and sticks there, distance becomes only 1/1,000mm, and radiation exposure will be 1 trillion times the exposure from a 1m distance.

distance between the substance and the cell is one millionth of 1 m. So the exposure dose, calculated as inversely proportional to the square of distance, will be one trillion times as much.

I explained above about the differences in penetration power of the radiation emitted from different substances. When radioactive substances are outside the human body, alpha or beta rays with poor penetration power hardly reach it. Only gamma rays can penetrate it. What monitoring equipment measures are gamma rays.

However, when internally exposed, the body will also be irradiated by alpha and beta rays. Plutonium, which is highly toxic and has long half-life, releases alpha rays. If it enters the human body, it will stay there and continue to subject it to irradiation. (figure 8).

After being internally exposed, the body may discharge the radioactive substance. However, genes may be damaged when chains of chromosomal DNA are cut while being internally exposed. If even one cell undergoes carcinogenesis, that is, changes into a cancer cell, it starts to proliferate. The human body is said to have secondary and even tertiary prevention mechanisms against cancer generation. However, there is no medical science that can prevent all DNA abnormality when undergoing continuous radiation exposure of multiple types.

Since radioactivity will damage genes in the process of cell division, the most vulnerable people are children and fetuses, whose cell division is very active, and whose and metabolic rate is very rapid.

Please take a look at figure 9 (page 66). It shows the changes in the frequency of thyroid cancer after the Chernobyl Accident. It shows that while there may be "no immediate effects" of irradiation, the effects will continue to surface slowly and over a long period of time. Five years after the accident, the number of occurrences spikes in the young generation. After 10 years have elapsed the trend seems to be subdued, but it only means that the affected people have shifted to the older generation, and (though the graph does not show this) the trend surges again almost 20 years after the accident. Now the affected generation has entered into the adult group. This is the generation which was exposed to radiation not only from the outside but also from the inside for a long time, and who develop cancer when they become adults.

Among radioactive substances, iodine 131 is carried easily by the wind and when it enters the human body, tends to concentrate in the thyroid glands

Figure 9
Thyroid cancer occurrence after Chernobyl Accident
(statistics from Belarus/ Minsk clinical cancer hospital)

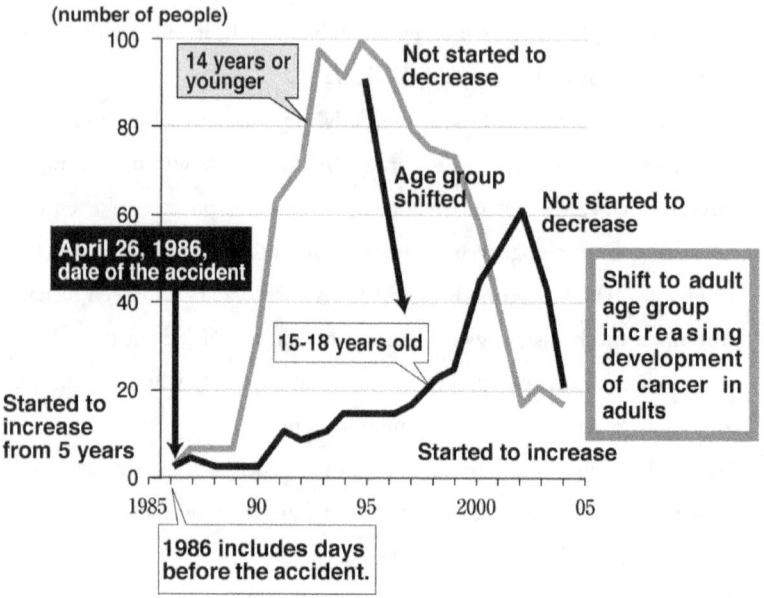

(number of people)

in the throat. Since iodine, necessary to produce thyroid hormone, is very
a important element for children and growing young people, the body
concentrates it in the throat without detecting that it is radioactive. Why
is radioactive iodine, with its short half-life of 8 days so dangerous for so
long? Because while the radioactivity of iodine is reduced by half every
8 days, which means that it will be reduced to a two-hundred-fiftieth in
almost two months, if the body is exposed directly to its radiation during
that two-month period, the damaged thyroid gland will, unbeknownst to
the person, begin to produce cancer cells. What is frightening is that these
deformed cells, far from being reduced by half each eight days, rather
increase rapidly in number.

Yazaki Katsuma (currently emeritus professor at Ryukyu University) ,who
has testified at a class action suit concerning the official recognition of

illness caused by atomic bombs, in his book Kakusareta Hibaku (Hidden Irradiation) (Shinnippon shuppan) wrote as follows:

"Outside Exposure [author's note: exposure from outside the body] is mainly by gamma rays, while internal exposure [author's note: exposure from inside the body] is mainly beta rays, but also gamma rays and alpha rays. Therefore, the exposure dose is much higher and its damage is more serious. The Japanese radiation scientists community is, in terms of internal exposure, the world's most obtuse."

Sawada Shoji, an emeritus professor at Nagoya University, submitted his opinion on the question of internal irradiation as a plaintiff's witness before a court adjudicating the question of official recognition of illness caused by the atomic bombs of Hiroshima and Nagasaki. At the time of the atomic bombing of Hiroshima he, aged 13 at the time, was at home 1.4km from ground zero, and suffered irradiation. What he says about internal exposure is incisive: "Measuring radiation inside human body is, as a practical matter, impossible. It should not be compared with external exposure." He also warns that people who exercise hard are particularly likely to intake radioactive substances into their body. He continues that the evacuation area should be expanded.

Giving Numerical Values to Radiation Has No Important Meaning

Whether human health is affected by internal exposure depends on not only the type and amount of radioactive substances that have entered the body but also on the constitution of the person. In considering the effect of exposure to radiation, whether internal or external, the individual should not be considered as having average features.

Individual differences are very large. In the case of blood pressure, for example, some people's is chronically high, while other people's rise only rarely. Without keeping a record of each person's blood pressure over their lifetime, you cannot judge what blood pressure is normal for this person, and what should be suspected as abnormal.

Sensitivity to radioactive substances and radiation also differs from person to person. I noticed when I was translating US / European medical documents that many healthcare professionals in the US and Europe are resourceful about such deviations among individuals, and are far more sensitive to the risk of radioactivity than Japanese medical service workers. Harada Masazumi (formerly of Kumamoto University) a medical doctor who has diagnosed and researched Minamata disease for a long time, told me "Get a firm understanding of personal differences and do not overlook patients who suffer without speaking out." Dr. Harada, who has observed Minamata disease from the patient's point of view, has had the painful experience of himself failing for a long time to notice many of the suffering people.

When Minamata disease symptoms started to surface among the people of Minamata, Dr. Harada was everywhere on the site diagnosing patients. At that time, he told me, he excluded people with innate poor health from his diagnosis. In reality, those were the people who were the first to fall victim to the disease. Dr. Harada made himself a scholar of personal differences and passed this warning on to me, knowing I am concerned with the effects of radiation.

I have the highest respect for Dr. Harada, and his words are engraved in my memory. And as I wrote above, I do not think it is proper to look at human health through simple numerical values.

For example, on March 22, 10 days after the Fukushima Daiichi accident, radioactive substances, which had been found earlier in the drinking water of Fukushima, were also found in the drinking water of Tokyo. At the Kanamachi Purification Plant in Katsushika-ku, Tokyo, 210 Becquerels per liter of iodine 131 were found. (Becquerel is the unit to indicate strength of radioactivity.) The intake limit for adults was fixed at 300 Becquerels per 1 liter of water and for infants it is 100. So the city of Tokyo called on the general public not to give tap water to infants in areas using water from the Kanamachi Purification Plant.

What surprised me at that time was the fact that infant intake limit was fixed at a third that of adults. This standard, which was set on March 17, almost a week after the earthquake and Fukushima meltdown accident, by the Nuclear Safety Commission (NSC) and the Ministry of Health, Labour and Welfare (MHLW) must not be trusted. In Europe and America, where people are more sensitive to radioactivity risks, setting the standard for infants at a tenth that of adults is the medical common sense. Still more surprisingly, in this provisional set of standards, the intake limitation for radioactive iodine 131 in vegetables is set at 2,000 Becquerels per kg. In the CODEX guidelines established by the UN Food and Agriculture Organization (FAO) and the World Health Organization (WHO), the limit is set at 100 Becquerels. In other words, Japan increased the permissible amount of iodine to 20 times the international standard. After this ridiculously high (dangerous) standard was established, TV and the newspapers have repeatedly reported that it is "a few times as much." Given this situation, even though radiation measurements yield figures far below the standard, children may be affected seriously and are highly likely to develop childhood cancer.

The government makes its case for safety using the notion of the threshold. However, below their threshold is an area where, as a matter of fact, as many as 10 percent of the people so exposed may develop disease. In a few years, it is certain that a large number of patients will appear. To begin with, when the Ministry of Health, Labour and Welfare set the standard referred to above, it based it on a standard set by the International Commission on Radiological Protection (ICRP). However that organization is dedicated to the promotion of nuclear weapons development and the nuclear power industry, and its idea of a safety standard is simply an arbitrarily selected figure, whose purpose is to put into everyone's heads the notion that "nuclear radiation below some particular level is safe." Similarly, those self-appointed "radioactivity researchers" who talk about safety on TV and in the newspapers and magazines are all bought by the nuclear power industry, to promote the effective use of radioactivity.

Will the marketers or manufacturers of agricultural chemicals say "agricultural chemicals are dangerous."? The same logic applies here. The people who talk about safety on TV or in newspapers and magazines are all cat's-paws of the nuclear power industry. I know this because I have investigated the background of each one of them. I will report on this in my next book that is forthcoming from Shueisha Shinsho. These people are very anxious now, because if "radiation is safe" turns out to be false, they will lose their source of income. Of course if anyone wants to believe that radiation, or agricultural chemicals, are safe, they are free to do so.

But there is no medical evidence to support the argument that radiation is safe because its numerical value falls below a certain safety standard. It is no great wonder if cancer develops below that standard. Except for unavoidable medical uses, human beings should not use radioactivity and radiation.

Evacuate West, Even If It Means Moving

Regarding the radiation leak, after the Fukushima Daiichi accident, many medical people have appeared in the media to make comments. Some doctors use the internet to make comments. Dr. Harada, who I mentioned above, said, "For doctors to make statements on environmental issues – it's too late now." I feel the same way.

From day one of the Fukushima meltdown, I warned, "a large amount of radiation is escaping into the atmosphere." Against this the government, NISA representatives and bought scholars who called themselves radiation experts appealed to the public through all TV stations available saying "Please be careful about rumors." "Please do not believe wild tales." And they repeated their "radioactivity is safe" story. What happened after that? As I mentioned in the preface of this book, a month after the accident, on 11 April, the Nuclear Safety Commission (NSC) announced that the Fukushima Daiichi Power Plant "in the first hours after the accident, was emitting as much as 10,000 terabecquerels of radiation per hour (one terabecquerel = one trillion becquerels – bq)", and acknowledged that the accident was at the same level as Chernobyl and was extremely dangerous. In effect, they admitted that it was the Kan Naoto Administration that had been spreading rumors and wild tales. For a government that has the duty to protect the health of the people, and most of all the health of children and young people, this is a woeful, venal, and irreversible crime

Highly radioactive substances have been detected in the seawater near the plant. Major ocean pollution is surely spreading, with radionuclides flowing south, riding the Oyashio current. Many thousands of tons of seawater – later, of fresh water – have been poured on the reactors to cool them down and keep them from blowing up entirely. In a nuclear plant, a jumble of countless ducts is located underground, carrying the wiring and plumbing. From the early stage of the cooling operation, I warned "They are not cooling it down. What they are doing is washing the reactor's radioactive substances and flushing them outside. The water is spurting out of all the many channels leading out of the facility." And that is in fact what was happening.

On April 4 they began discharging large amounts of irradiated water into the sea, saying that there was no other way to secure a place to pool highly polluted water, which infuriated the world community. TEPCO used the term "low level contaminated water". Ignorant to the last, TV and newspapers called it "slightly contaminated water", misinforming the Japanese people. However, the radioactive substances contained in this water are the same as those contained in high-level contaminated water; only the concentration is lower. And in fact the level is not low; it is 500 times the safety standard. If someone asked you to drink watered-down cyanide, would you do it? They are putting poisons like that, in an enormous scale, into the sea as though it were nothing. For us Japanese who, unlike Americans and Europeans, take 40 % of our animal protein from seafood, it is as though we were digging our own grave.

On April 12, radioactive strontium, which has properties similar to those of calcium, was detected outside the 30kg evacuation area. It is frightening to think about the route it must have followed, riding on the wind. For strontium 90 to accumulate in the bones of growing children, where it continuously irradiates them from the inside and causes leukemia, is something we absolutely must avoid. Strontium 90 has a half-life of around 29 years, and releases beta rays which are especially dangerous inside the human body. When large amounts of radioactive substances are released from a nuclear power plant, they fall eventually to the ground, concentrate there, are washed by the rain into rivers and from there into irrigation waters or the ocean. The water will be consumed by plants and animals, which means that the radioactive materials have entered a longer contamination system, the food chain, where they are further concentrated over a long period of time. Looking at graphs showing the amount of radiation in the air going down day by day, many Japanese feel relieved, but this is a total misunderstanding. Even though the amount of radioactive substances released into the atmosphere is reported to be decreasing, the issue is not the amount released daily but how much has accumulated in water and soil over the whole time period. From the day of the first hydrogen explosion, the absolute amount of radioactive substances which have escaped into the atmosphere and accumulated, has been increasing

Figure 10
Radioactivity concentration data taken from the Columbia River, USA

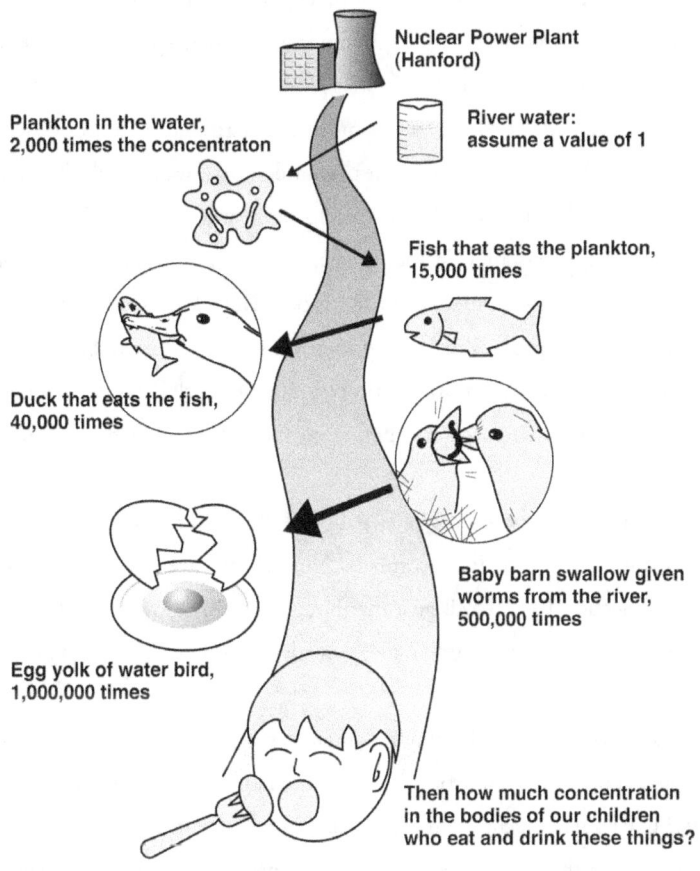

Nuclear Power Plant (Hanford)

River water: assume a value of 1

Plankton in the water, 2,000 times the concentraton

Fish that eats the plankton, 15,000 times

Duck that eats the fish, 40,000 times

Baby barn swallow given worms from the river, 500,000 times

Egg yolk of water bird, 1,000,000 times

Then how much concentration in the bodies of our children who eat and drink these things?

steadily. The thing to fear is the accumulation of those substances inside human bodies.

Please look at figure 10. This is a data collected by a scientist who studied the Columbia River in the USA.

Upstream, a nuclear power plant called the Hanford Reprocessing Plant releases diluted, low level of radionuclides into the river. Suppose the radioactivity of the substance to have a numerical value of one. It will be

2,000 times more concentrated when it is eaten by plankton. When the plankton is eaten by a fish, it will be 15,000 times more concentrated. If the fish are eaten by a duck, it will be concentrated to 40,000 times the original radioactivity. In this way, radioactive substances are concentrated in the food chain in nature.

Furthermore, it is 500,000 times more concentrated in the body of a baby barn swallow. When it comes to an egg of a water bird, the figure is 1,000,000. Therefore, it is not true to say that it is safe if the level of radioactive concentration in the water is low. The radioactive substance will be increasingly condensed in the food chain. We experienced a real example of this process in Japan during the world's worst pollution case, Minamata disease. And now, the process of radionuclide concentration in the food chain is under way, centering on the Fukushima Daiichi Nuclear Power Plant.

The radioactivity pollution caused by the Fukushima genpatsu shinsai syndrome has spread far too much. What should we do? Here, at the end of Section One, I will give my thoughts at this moment. These are my own, completely arbitrary ideas; readers will have to use their own judgment.

People under 30 years old, children, infants, pregnant women and younger women should evacuate from a zone with a radius of at least 250km from the Fukushima Daiichi Plant. They should move as far away as possible, and consider the move to be permanent. Considering the annual prevailing winds, western Japan should be safest in the long run.

The government should make all the preparations needed to receive these evacuees. Nationwide cooperation will be necessary. TEPCO has the responsibility to put its full energy into this effort.

Measurement of the contamination of food should be continued over a long period of time. This data should not be concealed but publicized, and all dangers should be made known to the people.

Individuals over 30 should assess the danger for themselves, and make a decision as to how to live their lives. Then, [if they so choose] they should remove all the shipping restrictions from contaminated food, and eat it. And they should drink the contaminated water. The government will go on telling them, "There will be no immediate effect on health." People who want to develop cancer in their later years should be free to do so.....

According to a radiation investigation performed inside Fukushima prefecture, as of the end of April, 75.9% of the elementary and middle schools showed radiation readings in excess of 0.6 microsieverts per hour (=5.3 milisieverts per year) which should qualify them as "radiation controlled areas" from which the public would be excluded. In other words, school children in Fukushima Prefecture are being put in the same circumstances as workers in a nuclear plant. The classrooms should be closed down and the children evacuated as soon as possible. I am praying that by the time this book is published, that will have been done.

As I explained in the preface to this book, panic occurs where people are not told the truth. The government, should understand this, and should inform people of the dangers, and of the countermeasures that are being taken. And, for the sake of our children, we must consider all possible measures. I don't care if I get cancer myself. But we have to protect the children. They are the ones who will decide the future of the country in 10, 20, or 50 years.

Nuclear Power Generation
and major Nuclear Power Plants in Japan

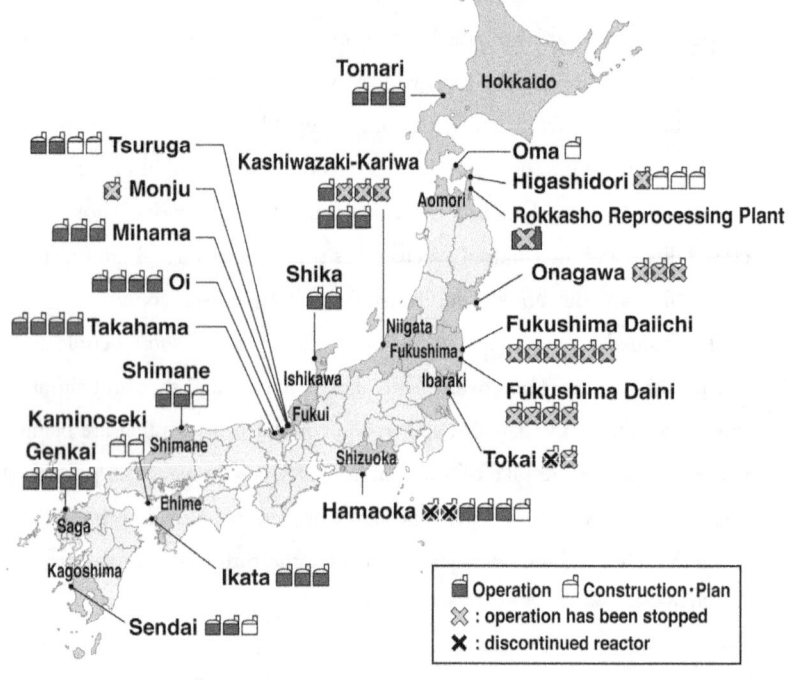

Chapter 4:
Japan Has Entered an Age of Major Earthquakes

The large-scale accident at the Fukushima Daiichi Nuclear Power Plant proved that genpatsu shinsai syndrome was not a desk theory; it has actually happened. My gloomy prediction came true although I absolutely did not want it to.

Many intelligent people, including seismologist Ishibashi Katsuhiko, had pointed out the danger of nuclear power plants. Nevertheless, no effective measures were taken. I can find no words to offer our children's generation but "I am sorry, forgive us." At the same time, I feel a new fear rising up within me.

The great fear is that there are many nuclear power plants in the Japanese archipelago that could become the 2nd or 3rd Fukushima. These nuclear plants could cause catastrophes exceeding the Fukushima disaster and thus affect the whole country and possibly the world.

As is known by people the world over, the Japanese islands are located on a prominent earthquake zone. And for these past ten-odd years, many Japanese people feel there has been an unusual succession of large-scale earthquakes.

In 1995, the Great Hanshin Earthquake occurred. In 2004, in Niigata Prefecture, the Chuetsu Earthquake occurred, badly damaging Yamakoshimura Village. Three years later, in 2007, the Chuetsu Offshore Earthquake shook the Kashiwazaki Kariwa Nuclear Power Plant, heavily damaging it. In 2008, the Iwate-Miyagi Inland Earthquake triggered a huge landslide, causing a whole mountain to disappear. Subsequently, the

Suruga Bay Earthquake in 2009 put the Hamaoka Nuclear Power Plant in Shizuoka Prefecture into the state of emergency. Now in 2011, the Great Tohoku Earthquake has occurred. From a geological point of view, several years means but a brief flash in time. We can say these quakes took place in rapid succession. Moreover, it is expected that this active phase will continue for some decades.

If so, then we need to understand that the Tokai Earthquake, which seismologists have been warning could come at any time, and whose hypocenter would be offshore from Omaezaki in Shizuoka Prefecture, and also the Nankai Earthquake, which has been predicted to occur offshore from Shikoku Island, near the Kii Peninsula, are on their way. [Translator's note: tokai means "eastern sea," and nankai means "southern sea." These two periodically recurring earthquakes are in effect the northeastern and southwestern versions of the same earthquake.] These earthquakes are expected to be in the magnitude 8 class, but the violence of nature can easily exceed man's arbitrary predictions, as the recent Great Tohoku Earthquake has taught us. We need to be prepared for the possibility of a magnitude 9 class earthquake striking the very center of the Japanese Archipelago. And directly on the epicenter of that earthquake is Omaezaki where Chubu Electric Power Company's Hamaoka Nuclear Power Plant is located.

And if you investigate the geological features and earthquake faults around Japan's other nuclear power plants, you will understand that the risk of earthquake is high at all of them. Even if it does not have as great a magnitude level as the Tokai or the Nankai Earthquakes are expected to have, an earthquake occurring directly below can cause terrible damage, as we learned from the Great Hanshin Earthquake of 1995. A genpatsu shinsai syndrome, as happened at Fukushima or even worse, could happen anywhere, any time, in Japan. If that happens, we can't get away with saying, "I didn't know that".

Before moving on with this chapter, there is one thing I would like the readers to have burned into their memory. That is the fact that among Japan's seismologists, the vast majority - 99% - of these "specialists" never once gave warning to the people of the danger of an accident at the Fukushima Nuclear Plant. Similarly, neither television nor newspapers warned the people of this danger. Please think over the fact that the number of genuine scientists who, like Professor Ishibashi Katsuhiko, risked everything to warn of the danger, was very small.

I say with certainty: "specialists" who keep their mouths shut are no specialists. If they insist on styling themselves as experts, they should speak out to protect people's lives. They should raise their voices for the abolition of nuclear power plants in Japan. That all through the Fukushima nuclear accident, and even now, no such voice is heard from them means that they are not to be trusted as human beings. Warnings mumbled in a small voice are not warnings.

Why Are There So Many Major Earthquakes in the Japanese Archipelago?

To begin with, why is it that major earthquakes occur so often in and around the Japanese Archipelago? I would like the readers to understand that, as the Archipelago is located in a region where there is no escaping from the danger of earthquakes, we who live here must build a society based on the assumption that earthquakes will come, and be prepared for them.

An earthquake is a movement of the ground, and the ground means Earth. To understand the mechanisms of earthquakes, it is necessary to know how Planet Earth is structured.

If you look at Figure 11 (page 80), you see that if you slice through the earth, its cross section looks something like that of an egg.

Figure 11
Structure of the Earth's Interior

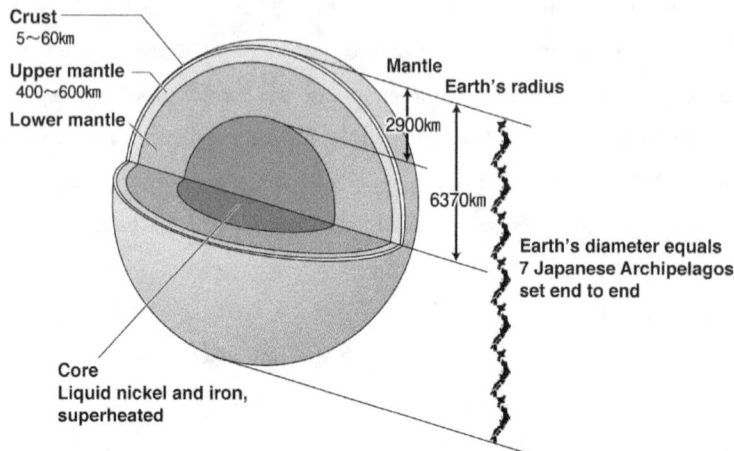

Crust
5~60km

Upper mantle
400~600km

Lower mantle

Mantle

Earth's radius

2900km

6370km

Earth's diameter equals
7 Japanese Archipelagos
set end to end

Core
Liquid nickel and iron,
superheated

The part that corresponds to the yolk of the egg is the core. It is believed that the core is made of molten metal at a temperature of several thousand degrees centigrade.

The white of the egg is the mantle. The mantle consists of high-density rock which, though solid, circulates like liquid water because of the high temperature at the center. This so-called mantle convection amounts to only a few centimeters a year, but if looked at from a geological time scale of millions of years, it can be said to be very active. If the mantle moves only five centimeters a year, that means 500 kilometers in 10 million years and 5000 kilometers in 100 million years.

Finally, the part that corresponds to the shell of egg is called the crust. It includes the surface where humans live and the bottom of the ocean. From man's point of view, the crust, at roughly 5 to 60 kilometers, seems very thick. However, it is thin compared to the Earth's diameter of 12000 kilometers. Rather than corresponding to the shell of the egg, it is more precise to say that it corresponds to the thin film that remains after the egg is peeled.

This film-like crust is not a seamless surface, but is divided into many plates. These plates, pulled along by the circulation of the mantle beneath them, move slowly in different directions.

Of course, we cannot see the structure of the Earth's interior. But studies of such things as seismic waves, geomagnetism, volcanic activities and the composition rocks have made it clear that the Earth acts as not as "solid ground" but more like a living thing. The plates, with which the surface of the Earth is covered, are constantly moving. And the phenomena caused by that are earthquakes.

The branch of geology which seeks to understand the structure of these plates is called plate tectonics.

On the Boundary of Four Plates

A hundred years ago, a German geophysicist, Alfred Wegener, noticed that the geographical features of the east coastline of South America exactly fit those of the African west coastline, and he proposed the theory of continental drift. He also studied fossils of extinct animals and plants found in Africa and South America and noticed many similar features between the two continents. He argued that the present continents were formed by an ancient supercontinent, later named Pangea, splitting and colliding over many years.

At first, few people accepted the theory of continental drift. But in 1929 the British Arthur Holmes proposed mantle convection theory, which supported the notion. Mantle convection theory, described above, asserts that it is the movement of the mantle that moves the continents. Then from the 1960's approaches such as the new Earth archeology advanced, producing new pieces of evidence one after another. Especially, geomagnetism was conclusive. A study of rocks from all over the world, in particular of the traces of geomagnetism left in them from the time they were originally formed, provided strong scientific evidence supporting

Wegener's theory.

Then in 1968, Canadian geophysicist Tuzo Wilson collected the various theories related to continental drift and organized them into the theory of plate tectonics. Plate tectonics is now the most important theory in seismology and geophysics.

Looked at from the standpoint of plate tectonics, how does the area around Japan appear?

Put simply, Japan is located on the boundaries between four plates, the Eurasian Plate, the North American Plate, the Philippine Sea Plate, and the Pacific Plate. The Pacific Plate, the biggest and fastest moving plate on Earth, has a strong influence on Japan. At its eastern boundary, the mantle's convection rises up from what is called a submarine ridge, actually a submarine mountain range that runs offshore along the coast of North and South America. At the western edge, the plate returns to the Earth's interior at the Japan Trench, off the Pacific coast of Japan. The Japanese Archipelago is located at an area where four plates meet, colliding and subducting with enormous power (subduction is the name for one plate forcing its way beneath another).

Offshore from Izu Peninsula in Shizuoka Prefecture, the Philippine Sea Plate is subducting under the Eurasian plate, and the Pacific Plate is driving beneath both of them at a depth of 1000 kilometers. Nowhere else in the world do three plates converge at one point in this dangerous manner.

Many volcanic regions, including Mt. Fuji, Hakone, Izu Oshima Island, and Miyakejima Island, are distributed around Izu Peninsula. Furthermore, the area is the epicenter from which major earthquakes on the scale of the Kanto or the Tokai earthquake could take place. The theory of plate tectonics indicates that tremendous energies strike against each other under the earth of this region.

The Japanese Archipelago: A Whole Era Younger

Where the four plates press against one another, the result is intense volcanic activity and orogenic movement, which is what created the Japanese Archipelago.

The lava and volcanic ash that gushed from the volcanoes accumulated over time to form stone. The activity by which the earth's crust is raised up to form dry land and mountain ranges is called "orogenic movement". The Japanese Archipelago was formed by this orogenic movement, but that doesn't mean that the movement has come to an end: it goes on forever, it is going on now, a fact that the Japanese people seem to have forgotten.

Originally, Sakurajima was an island off Kagoshima Prefecture, in the southern part of Kyushu. In 1914, a tremendous volcanic eruption took place on Sakurajima. The flowing lava connected it to the mainland. In the years between 1943 to 1945, on a plain in Hokkaido, a new volcano created a new mountain, Showa Shinzan. In 1973 a submarine volcano erupted suddenly near Nishinoshima Island in the Ogasawara Island chain. A new island, Nishinoshima Shinto was formed within half a year. Volcanic activity continued for two years, and the new island ultimately connected with the existing Nishinoshima Island. Thus, even today, the volcanic activity is causing dynamic geographical changes in the Japanese Archipelago. In January 2011, the Shinmoedake volcano in the Kirishima mountain range in Kyushu erupted. It was shortly after that, March 11, that the devastating Great Tohoku Earthquake occurred.

It is believed that the Japanese Archipelago was given its present form, separated from the continent by the Japan Sea, about 30 million years ago. 20 million years before that the Median Tectonic Line, which runs from northern Kyushu through central Japan and Shikoku Island as far as north-central Honshu, and is one of the largest active faults in the world, already existed. From some 2 million years ago, the upheavals that raised up the Japanese Archipelago have been continuing. From about 1 million years ago the great rift valley called Fossa Magna, which passes under central Japan from the Pacific Coast toward the Japan sea, began to form Japan's Northern Alps (Hida Mountains). With this massive orogenic movement and with pressure coming in from four directions, the power formation beneath Japan became very complex, with many cracks and faults. It is because this is the way the Archipelago was formed that it is our fate to experience frequent earthquakes.

Moreover, after the islands finally took their basic form, there came recurrent ice ages and interglacial periods, bringing radical climactic variation to the newly formed islands. During the ice ages, the temperature of the entire Earth fell, glaciers were formed, and the sea level dropped. As a result the continent was connected by land with the Japanese islands, and people and animals could walk across it. During the clement interglacial periods, the ice melted and the sea level rose. For a long period, much of the present coastline was under the sea. Therefore, shell mounds made by people during the Jomon Period (16500-3000 years ago) have been discovered even in the recesses of inland mountains, for example in Ibaraki Prefecture.

What this means is that much of Japan's present coastline was at the bottom of the sea just a few thousand years ago. As a result, many of Japan's coastal areas are not formed of solid bedrock, but are fragile ground built up of dirt and mud. However, almost all of Japan's nuclear power plants, because they need to use seawater for cooling, are constructed on this fragile coastline.(Chapter 6 explains in detail the location of each nuclear power plant.)

The geological features of Europe are different. The ground on which Finland and Sweden sit is old, formed some two billion years ago. The land north of the Alps was formed between 500 million and one billion years ago. Moreover, though around the Mediterranean Basin, where for example in Italy and especially Sicily volcanoes are frequent, the land is relatively new, still 200 million years or more have passed. On the other hand, until about 3000 years ago, the Japanese coastline was still shifting, sometimes sea bottom and sometimes land. We need to realize that the recent formation of the Japanese Archipelago compared to other regions amounts to a categorical difference.

The Inter-plate Earthquake and the Inland Direct Hit Earthquake

As we have seen, the Japanese Archipelago is geologically very unstable, and world-class earthquakes frequently occur. What, then, is the mechanism that causes them?

There are various detailed seismological classifications describing earthquakes. But there are two main categories, the inter-plate earthquake and the inland-direct hit earthquake [translator's note: the Japanese expression for the latter is nairiku chokkagata jishin. Translated literally, this would be "inland directly-below-type earthquake." The technical expression in English is "near field earthquake". Hopefully the more popular expression "inland direct hit" will communicate better to the reader the nature of this earthquake].

The inter-plate earthquake is, as the name implies, caused by the movement of the very plates of Plate Tectonics, that is, by the movement of the slabs of rock that make up the crust of the earth. Western Japan is on the Eurasian Plate, Eastern Japan is on the North American Plate, the Philippine Sea Plate spreads from the south and the Pacific Plate moves under it from the east. Where the edge of that plate drives into the earth the bottom of the sea is cut deep; one such deep, off Japan's Pacific shore, is called the Japan Trench; another, which extends from the coast of Shizuoka to the coast of Shikoku, is called the Nankai Trough. (At the bottom of the sea, a slash deeper than 6000 meters in depth is called a "trench"; a shallower one is called a "trough".)

The edges of the Eurasian Plate and the North American Plate are dragged down when the Pacific Plate and the Philippine Sea Plate slowly slide under them. They spring up again when the distortion accumulated there exceeds a certain limit; this energy causes earthquakes.

The Great Tohoku Earthquake that occurred in March is an exact model of the inter-plate earthquake mechanism. The impending Tokai and Nankai Earthquakes will be of the same type. Since the inter-plate earthquakes occur around a trench or trough, where the plates meet, it is also called a "trench type earthquake".

The characteristics of the inter-plate earthquake are that it is very large, and that it occurs periodically. It is the mantle convection moving the plates, that is, it is global scale energy that is its cause, so it is not unusual for the resulting earthquakes to be of the magnitude 8 class. And because the speed at which the earth's crust moves and the limit to the amount of stress a plate can tolerate are both fairly constant, these earthquakes occur at pretty regular intervals (Figure 12 on page 88).

When one place somewhere along a plate springs up, earthquakes are caused one after another. In the Great Tohoku Earthquake, it is thought that at least three places under the sea from Iwate Prefecture to Ibaraki Prefecture sprang up in rapid succession, which was the reason for its great scale. And historically, the Tokai and the Nankai Earthquakes have always followed one another in a very short time.

When it occurs offshore, an inter-plate earthquake creates a tsunami. Produced when a section of the ocean floor hundreds of kilometers in length suddenly rises up several meters, the power of the tsunami needs no introduction. As I mentioned in Chapter 1, both the tsunami that followed the 1896 Meiji Sanriku Earthquake and the tsunami that came this year reach a height of 38 meters.

On the other hand, inland direct hit earthquakes are caused by cracks in the bedrock directly beneath the Archipelago. In an area which has been laced with cracks and faults from past orogenic movement, now complex pressure comes in from all sides by four moving plates, which produces distortions everywhere. When this distortion reaches a critical limit, the Earth's crust fractures and the energy released causes earthquakes.

While the inter-plate earthquake occurs at regular intervals in the vicinity of a trench or a trough, it cannot be predicted where or when an inland-direct hit earthquake will occur. However, at a place where a large movement exposed a major fault in the past, movement on that fault might occur again if enough distortion energy is collected. Therefore, if the fault signs indicate that movement took place since 1.7 or 1.8 million years ago seismologists call it an "active fault", and consider it a danger area. However, ruptures in the bedrock are not easily found, and there are many examples in which, even after a major earthquake has occurred, research at the epicenter cannot discover the fault. Thus the presence or absence of an active fault cannot by itself predict the danger of earthquake.

The inland-direct hit earthquake, when compared to the inter-plate earthquake, does not have greater energy at the epicenter, but as it takes place directly below where people live, it can produce stronger shocks there than the inter-plate earthquake, and can cause great damage.

The Great Hanshin Earthquake, which was an inland-direct hit earthquake, had a magnitude of 7.3, less than one tenth that of inter-plate earthquakes such as the Tokai and Nankai earthquakes, but it had tremendous destructive power, knocking over expressways. In 2008, the Iwate-Miyagi Inland Earthquake hit in an area where it was not known an active fault existed; it caused an area two kilometers in radius to cave in, and triggered landslides to the extent that an entire mountain disappeared. The inland-direct hit earthquake presents a different danger than does the

Figure 12
Significant earthquakes around the Pacific plate,
2009-2010 (Dates in Japanese local time)

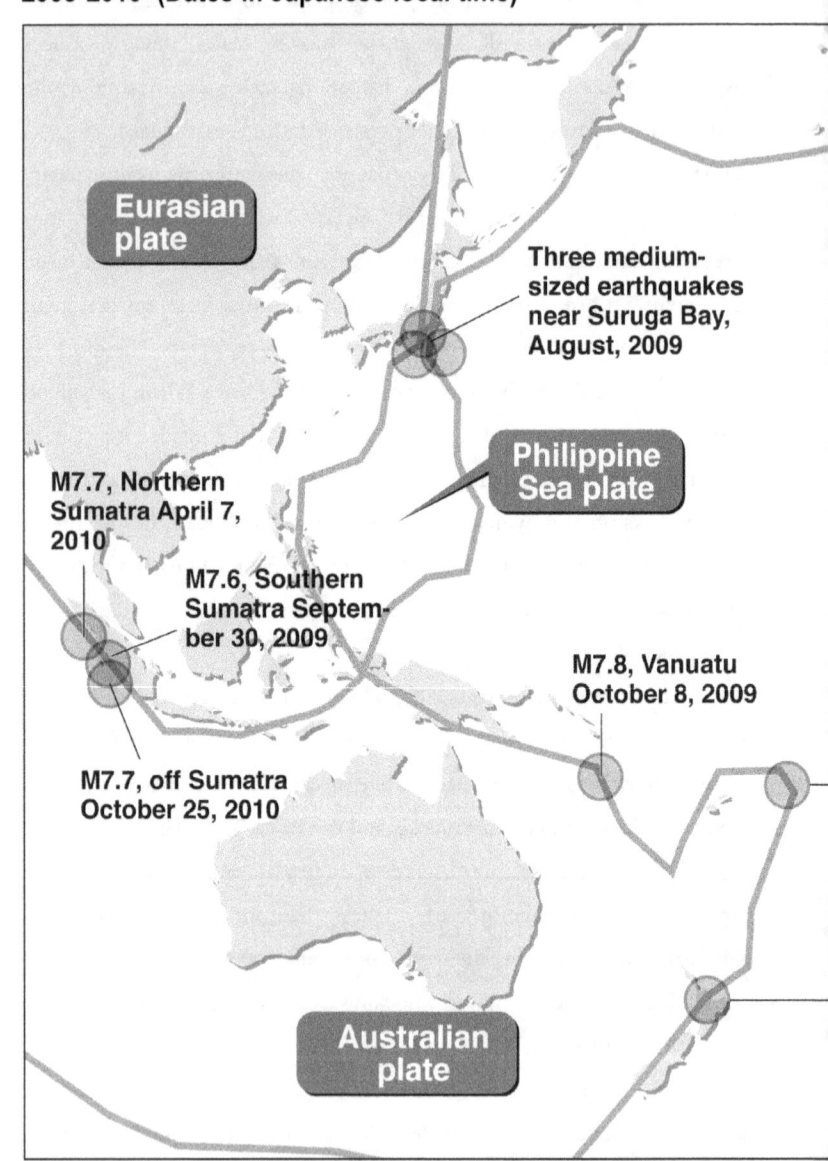

Eurasian plate

Three medium-sized earthquakes near Suruga Bay, August, 2009

Philippine Sea plate

M7.7, Northern Sumatra April 7, 2010

M7.6, Southern Sumatra September 30, 2009

M7.8, Vanuatu October 8, 2009

M7.7, off Sumatra October 25, 2010

Australian plate

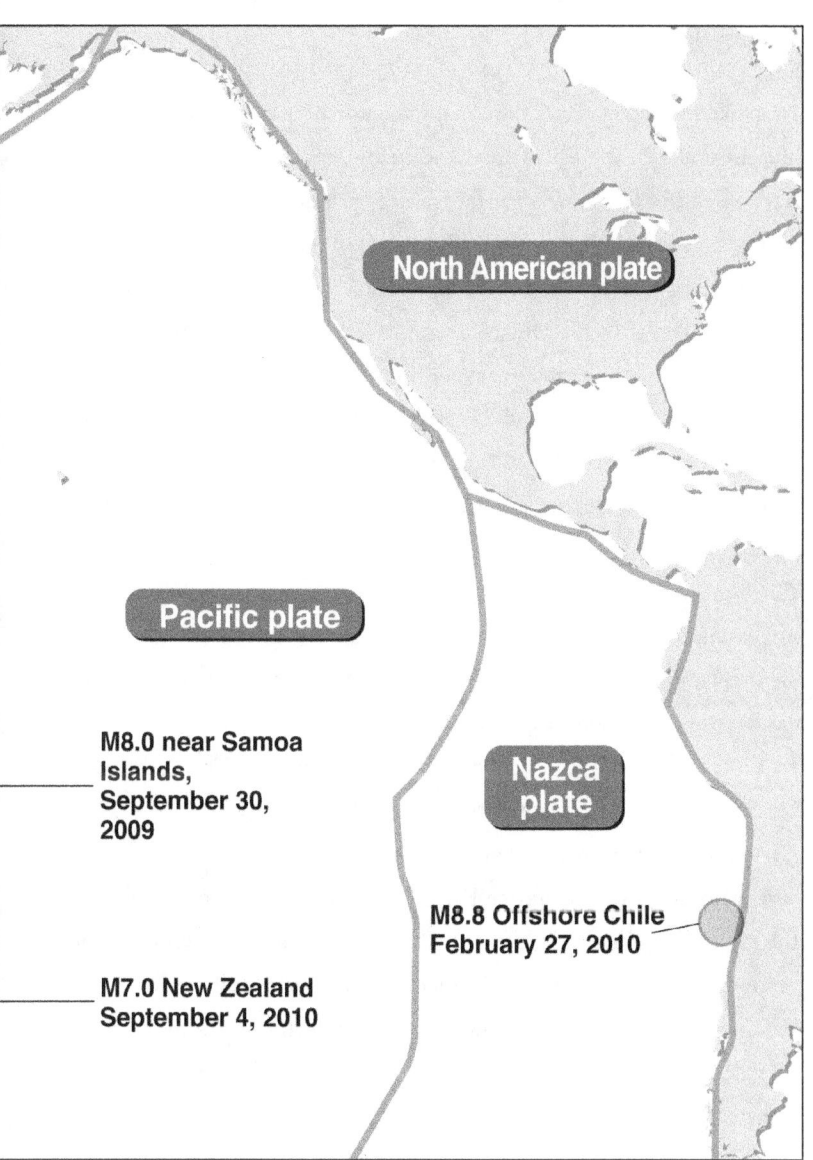

North American plate

Pacific plate

Nazca plate

M8.0 near Samoa Islands, September 30, 2009

M8.8 Offshore Chile February 27, 2010

M7.0 New Zealand September 4, 2010

inter-plate earthquake.

Frequent Small- and Medium-sized Earthquakes Harbinger a Major Earthquake

In quake-prone Japan, we have many records of past earthquakes from ancient times. In the Edo Period and before, though the people did not have scientific techniques for measuring earthquakes, they left written records, from which it is possible to estimate the *shindo* at specific locations, the epicenter, and the magnitude of the various earthquakes.

Professor Ishibashi Katsuhiko's historical investigation established that major earthquakes like the Tokai and the Nankai have been occurring at intervals of between 100 and 250 years, and since the 1300s, at intervals of less than 150 years. These facts fit with theory of plate tectonics, and constitute evidence that these major earthquakes were of the inter-plate type.

Earthquake records reveal another fact. These major earthquakes do not come unannounced, but rather both before and after they strike, almost without fail, there has been a number of medium scale or above earthquakes, and also volcanic eruptions, in the vicinity.

In 1707, in the early Edo Period, the Tokai and the Nankai Earthquakes hit one after another; the event was called by the collective name, Hoei Earthquake. It is recorded that tens of thousands of people died. Four years prior to that the Genroku Earthquake occurred, destroying most of the towns along the post road between Kawasaki and Odawara. In the same year as the Hoei earthquake there was the Hoei Eruption of Mount Fuji, which covered what is now Shizuoka Prefecture with three meters of volcanic ash. What kind of world would it be, under three meters of volcanic ash?

In the 1700's there was also the Great Eruption of Sakurajima, of Asamayama, of Unsenfugendake – volcanoes exploding one after another.

Bad weather caused by the Asamayama eruption caused crops to fail, causing what is known as the Great Tenmei Famine.

In 1854, the latter part of the Edo Era, the Ansei Tokai Earthquake, and the Ansei Nankai Earthquake struck as a set, and in a short space of 11 years before and after that there came, one after the other, the Zenkoji Earthquake, the Kaei Odawara Earthquake, the Ansei Iga Ueno Earthquake, the Ansei Edo Earthquake, and the Hietsu Earthquake. Though the population was much fewer then than now, still several thousand people died in each of these earthquakes, which makes them major disasters. And it is said that in the Tokai/Nankai Earthquake, the quake and the following tsunami killed 10,000 people in the Tokaido region, on the Pacific Coast south of Tokyo.

For forty years before the Great Kanto Earthquake struck in 1923, small and medium-sized scale earthquakes happened frequently over the Kanto region. In 1922, just half a year before the Kanto quake, which was magnitude 7.9, the Uraga Channel Earthquake hit, at magnitude 6.8.

When the pattern of occurrence of past earthquakes is examined, it can be seen that massive earthquakes such as Tokai and Nankai occur during a "tempestuous period" in the Japanese Archipelago, in which many earthquakes and volcanic eruptions take place. (Figure 13 on page 92).

Moreover, it is a particular characteristic of these harbinger-earthquakes and volcanic eruptions, that most of them occur along Japan's largest active fault, the Median Tectonic Line.

Today, the Japanese Archipelago is in a Tempestuous Period

In Japan, earthquakes and volcanic eruptions have been active for tensome years. If you list up the principal disturbances, it looks like the following.

Figure 13 Tokai and Nankai earthquakes occur at a constant cycle

157 years have passed since the Ansei Tokai earthquake; the next one is overdue.

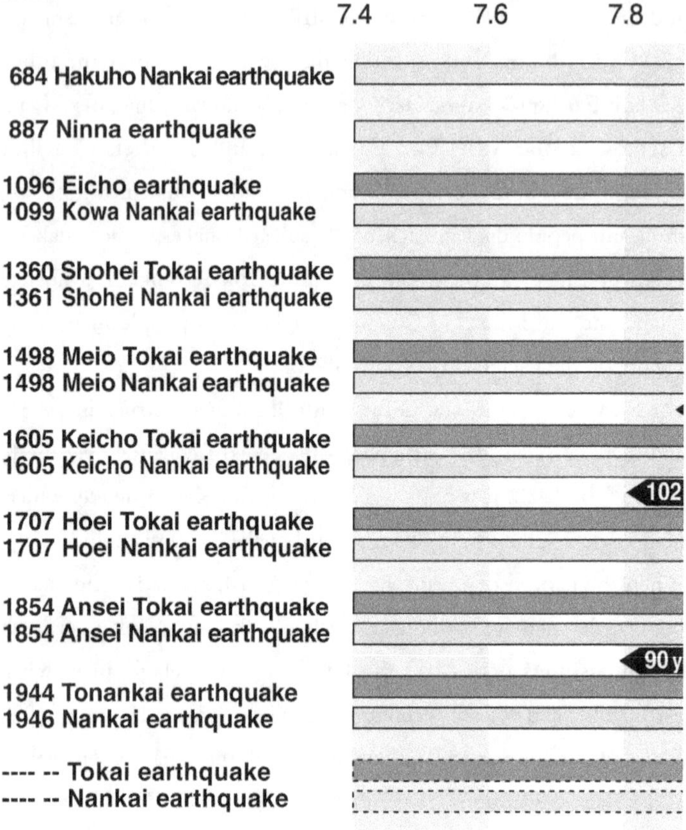

| | 7.4 | 7.6 | 7.8 |

684 Hakuho Nankai earthquake

887 Ninna earthquake

1096 Eicho earthquake
1099 Kowa Nankai earthquake

1360 Shohei Tokai earthquake
1361 Shohei Nankai earthquake

1498 Meio Tokai earthquake
1498 Meio Nankai earthquake

1605 Keicho Tokai earthquake
1605 Keicho Nankai earthquake

◀102

1707 Hoei Tokai earthquake
1707 Hoei Nankai earthquake

1854 Ansei Tokai earthquake
1854 Ansei Nankai earthquake

◀90 y

1944 Tonankai earthquake
1946 Nankai earthquake

---- -- Tokai earthquake
---- -- Nankai earthquake

1995   magnitude 7.3   Great Hanshin Earthquake, Hyogo Pref.

1997   magnitude 6.8   Sendai Earthquake, Kagoshima Pref.

2000   Great Eruption of Miyakejima, Izu Islands

2004   magnitude 6.8   Chuetsu Earthquake, Niigata Pref.

2004   Eruption of Asamayama, Nagano Pref.

2007   magnitude 6.8   Chuetsu Offshore Earthquake, Niigata Pref.

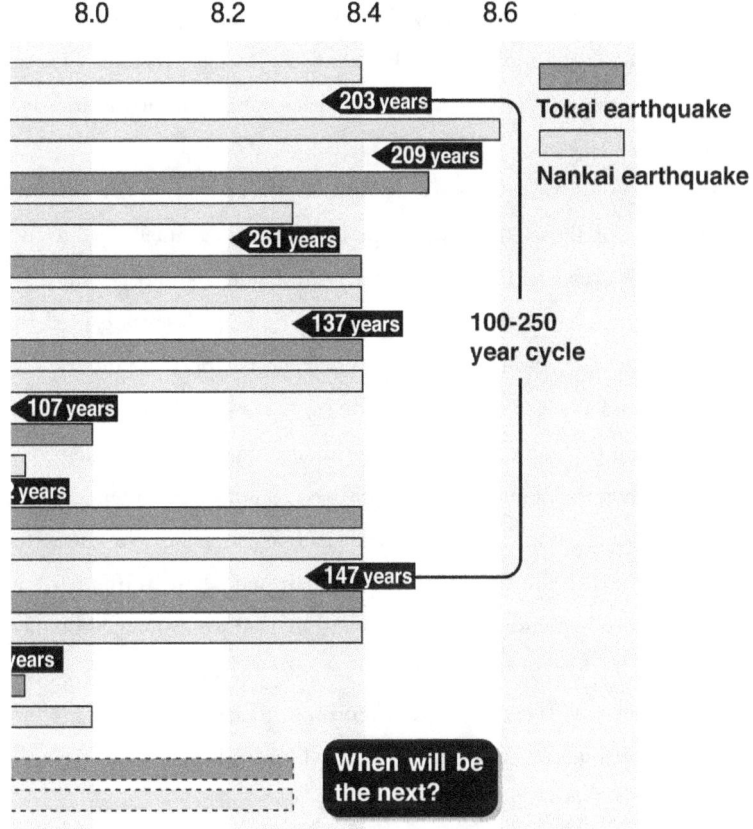

2008 magnitude 7.2 Iwate-Miyagi Nairiku Earthquake, Iwate and Miyagi Pref.

2009 magnitude 6.5 Suruga Bay Earthquake, Shizuoka Pref.

2009 Eruption of Sakurajima, Kagoshima Pref.

2011 Eruption of Shinmoedake, Kagoshima Pref.

2011 Great Tohoku Earthquake

The Great Eruption of Unzenfugendake, which occurred in 1991 and caused a huge pyroclastic flow (a rapidly moving current of superheated gas and rock) could, I think, be included in this list. Although it took place 20 years ago, on a geological time scale, 20 years is but a moment.

Excluding the two earthquakes that hit in the northeast, all of these seismic phenomena took place either along the Median Tectonic Line or the Kashiwazaki/Chiba Tectonic Line, which is located at the eastern edge of the Fossa Magna.

Considering these things, there is no escaping the conclusion that we have entered one of those "tempestuous periods" which eventually lead to the return of the Tokai and Nankai Earthquakes. (Figure 14 on page 96).

Moreover, it is evident the Great Tohoku Earthquake was caused by the movement of the Pacific Plate. When the Pacific Plate moves, the Philippine Sea Plate, which rides on top of it, is moved, so the Eurasian Plate, on top of both of them, will not remain quiet. As the three plates clash together, their influence inexorably bears on the hypocenter, invisible to human beings, from which the Nankai and Tokai Earthquakes originate. Four days after the Great Tohoku Earthquake, on March 15, an earthquake of magnitude 6+ occurred at Fujinomiya, in Shizuoka Prefecture. The epicenter was at the foot of Mt. Fuji, 5 km SSE of the summit, between the 3rd and 4th stations on the climbing route. Of the direct hit earthquakes under Mt. Fuji of which we have record, this was the strongest. Since then the region has been experiencing seismic swarm (many earthquakes striking in a short period), believed to be caused by the pressure of the magma, which some people take to be the sign of an impending major eruption. However, the government's Coordinating Committee for the Prediction of Volcanic Eruption of Japan considers volcanoes and earthquakes as separate phenomena, and the Earthquake Research Committee (located in a different branch of the government) announced that the rash of earthquakes on Mt. Fuji was irrelevant to the Tokai Earthquake. In the past the Tokai Earthquake has always followed a string of medium range earthquakes, so I can't comprehend the sensibility of these self-styled "specialists" to call it "irrelevant. Not once have we ever been given advance warning of a major earthquake or volcanic eruption by these irresponsible people.

Now in 2011, 157 years have passed since the Ansei Tokai Earthquake. As I explained, the Tokai and Nankai Earthquakes occur in a 150 year cycle. The massive earthquake time bomb has exceeded its time limit. For almost 50 years after the war, we experienced a quiet interval without noticing it. But with the Great Hanshin Earthquake, Japan entered "tempestuous period" of earthquakes; we need to engrave that fact into our understanding.

Refer once again to the list of earthquakes and eruptions that have occurred in these ten years.

Near the epicenter of the 1997 Sendai Earthquake in Kagoshima Prefecture, Kyushu Electric Power Company's Sendai Nuclear Plant is located. Near the epicenter of the 2004 Chuetsu and the 2007 Chuetsu Offshore Earthquakes in Niigata Prefecture, TEPCO's Kashiwazaki Kariwa Nuclear Power Plant is located. The 2009 Suruga Bay Earthquake put Chubu Electric Power Company's Hamaoka Nuclear Power Plant out of operation. It is as if the earthquakes were aiming at the power plants.

When hit by these earthquakes, the Kashiwazaki Kariwa and the Hamaoka Nuclear Power Plants went into automatic shutdown mode, and could not be restored to operation for a long time. But that these earthquakes did not hit the plants directly and trigger a genpatsu shinsai syndrome, was a matter of luck only. However the bad luck of being so lucky could not continue, and on 3/11 earthquake and tsunami overwhelmed the Fukushima Daiichi Nuclear Power Plant, and the genpatsu shinsai syndrome became a reality.

With the Archipelago now well into its "tempestuous period" of

Figure 14
The succession of abnormal phenomena in the last 20 years, what does it tell us?

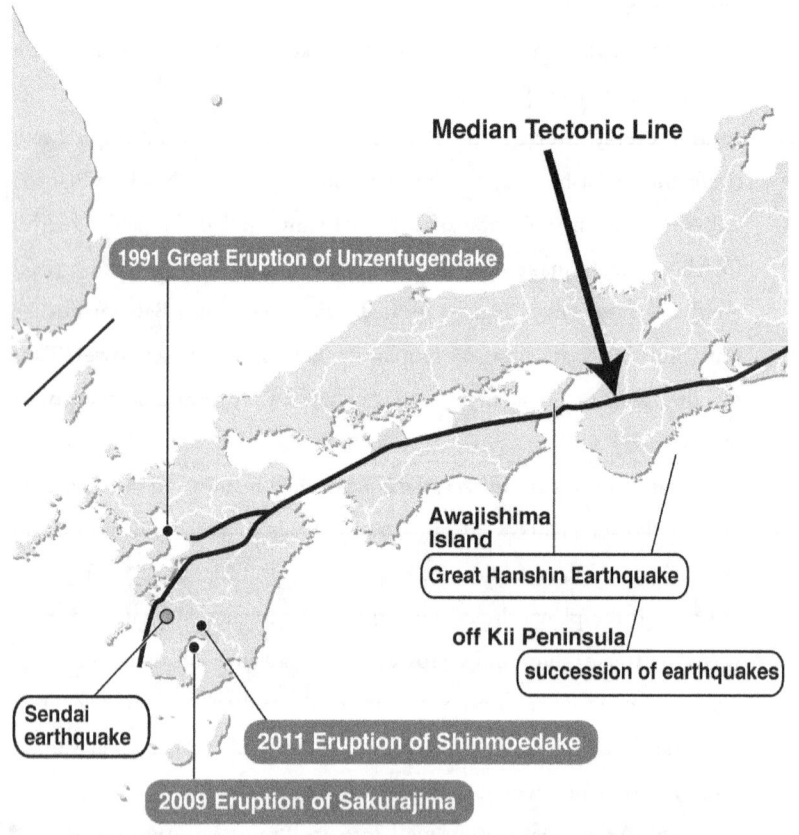

Median Tectonic Line

1991 Great Eruption of Unzenfugendake

Awajishima Island

Great Hanshin Earthquake

off Kii Peninsula

succession of earthquakes

Sendai earthquake

2011 Eruption of Shinmoedake

2009 Eruption of Sakurajima

earthquakes, it is a mistake to suppose that the Fukushima Daiichi Nuclear Disaster will be the last genpatsu shinsai syndrome we shall experience. Rather, the odds are there will be more. The leading candidate for the next is the Hamaoka Nuclear Power Plant.

In Chapter 5, we will look at the grave dangers facing the Hamaoka plant.

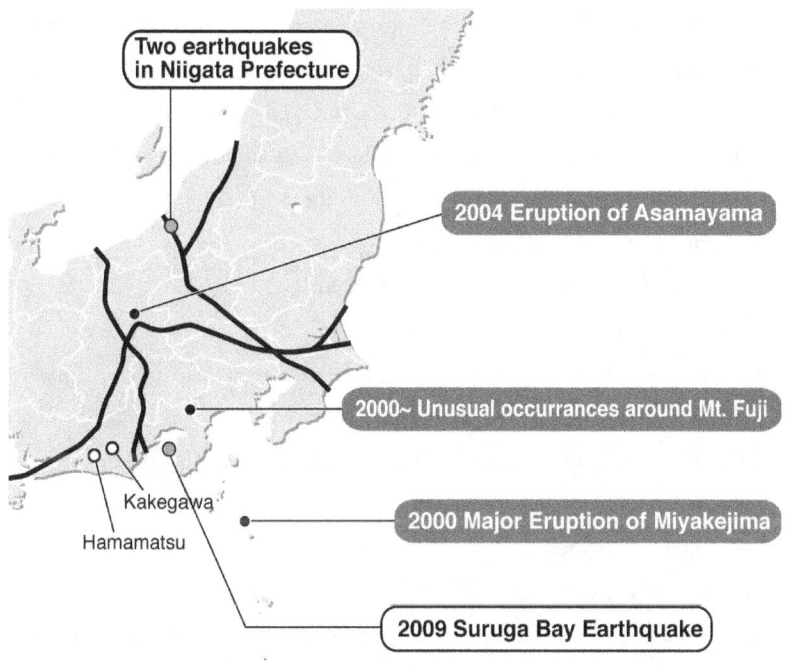

Two earthquakes in Niigata Prefecture

2004 Eruption of Asamayama

2000~ Unusual occurrances around Mt. Fuji

2000 Major Eruption of Miyakejima

Kakegawa

Hamamatsu

2009 Suruga Bay Earthquake

quakes)

The expected Tokai and Nankai Earthquakes will occur around the Nankai Trough extending from Suruga Bay to the south of Shikoku at the depth of 4000 meters, but many earthquakes and eruptions have occurred along the Median Tectonic Line over the last 20 years.

Chapter 5:
The Hamaoka Nuclear Plant: An Impending Catastrophe

At 10:31 AM on March 15th, 2011, just four days after the Great Tohoku Earthquake, another earthquake hit, shaking a large area of Honshu. Its epicenter was in eastern Shizuoka Prefecture and it had a magnitude of 6.4. In Tokyo, where I live, the trembling was ominously strong, and when I learned that the center was not in the northeast but in Shizuoka Prefecture, chills went down my back.

Is the Hamaoka Nuclear Power Plant all right? '

I couldn't help imagining a nightmare scenario: if the Fukushima accident were followed by an accident on the same scale at Hamaoka, the capital area of greater Tokyo would be trapped in between.

I think we are very fortunate that there has been no big earthquake in Shizuoka since then. But as I mentioned in Chapter 4, the record of the past shows that the possibility of a `Great Tokai Earthquake' is high after a middle-sized earthquake has occurred, so with this earthquake the danger has not lessened, but rather increased. The 2011 Great Tohoku Earthquake was caused by a shift in the earth's crust, at the boundary between the Pacific Plate and the North American Plate, on a scale very rare in recorded history. It is impossible that such a massive release of energy could have no effect on the nearby Philippine Sea Plate and Eurasian Plate. Now more than ever, we need to make preparations for the coming of a Tokai Earthquake.

Which raises the question, how can we prevent a Tokai Earthquake from destroying the Hamaoka Nuclear Power Plant and causing the next genpatsu shinsai syndrome?

The Tokai Earthquake: Directly Beneath the Reactors

As it turned out, the first actual case of a genpatsu shinsai syndrome tragedy took the form of the meltdown at the Fukushima Daiichi Nuclear Power Plant. But before that happened, it had been predicted that Chubu Electric Company (CHUDEN)'s Hamaoka Nuclear Power Plant located a Omaezaki, Shizuoka Prefecture would most likely be the first. Directly before the plant Suruga Bay spreads out; directly behind it pass the Tokaido bullet trains and Tomei superhighway, forming a main artery for transport in Japan. The plant is located halfway between the Tokyo capital and the Chubu area. The plant was built to provide energy to the Chubu economic region centered on Nagoya, which includes the Toyota automobile plant among other major industries. Reactors #1 and 2 of the Hamaoka Plant, which were built in the1970's, were shut down in January 2009 by reason of decrepitude and insufficient earthquake preparation and are ready for decommissioning. (This because Chubu Electric lost a court case filed by the citizens of Shizuoka Pref. demanding the shutdown of those reactors.) Now only reactors #3, 4 and 5 are operating. Among them, reactor #5 can produce 1.38 million kW and is the biggest nuclear reactor in Japan. However, after the big Suruga Earthquake on 11 August, 2009 (I will say more about this below), whose magnitude of 6.5 was beyond the company's expectation, it was temporarily shut down. But on 25 January, 2011, despite the fact that it was known as a faulty reactor, it was recklessly switched on for tune-up operations, and began generating electricity again on the 28th. It is difficult to believe.

The reason people think the Hamaoka nuclear power plant is dangerous is that its reactors sit directly on top of the fault that causes the Tokai

Earthquake, which occurs roughly every 150 years.

The Philippine Sea Plate slides gradually under the Eurasian Plate, pulling its edge down with it. When the pressure of this deformation reaches its limit, the edge springs back and causes a strong plate-boundary earthquake. This is the mechanism of the Tokai Earthquake. This earthquake, which is a typical plate-boundary earthquake, can, if it occurs, be expected easily to exceed a magnitude of 8. Moreover, the edges of these plates are not cut neatly like the edges of a jigsaw puzzle. The Philippine Sea Plate slides under the Eurasian Plate, and the area where they overlap is quite wide. And it is where they overlap that earthquakes can occur. The government's Headquarters for Earthquake Research Promotion predicts that the probability of the Tokai Earthquake occurring within the next 30 years is 87%. But as historically it has occurred in 150-year cycles, and the last one occurred 157 years ago, it is past due, and so could come at any time. Or rather, as the movement of the Pacific Plate during the 3/11 earthquake could not but have stimulated the Philippine Sea Plate, we can assume that the next Tokai Earthquake is on its way.

The overlapping boundaries of the Philippine Sea Plate and the Eurasian Plate lie beneath the Suruga Trough and the Nankai Trough which extend from Suruga Bay to Omaezaki Cape. And this active fault passes under Omaezaki, on top of which the Hamaoka Nuclear Power Plant stands. It means that the Tokai Earthquake is dangerous both for its scale, and for the fact that it is expected to occur directly beneath the Hamaoka Power Plant.

The 2011 Great Tohoku Earthquake occurred under the sea, offshore from Japan's northeast coast. The Tokai Earthquake, expected to be on the same scale, will extend inland, right under the Hamaoka Plant.

Nuclear Fission in Uranium Cannot Be Stopped

We have already experienced the destructive power of an inland earthquake with the Great Hanshin Earthquake. Elevated superhighways were knocked down by this earthquake, which had a magnitude of 7.3.

Supposing that a Tokai Earthquake occurred at a magnitude of 8.3; that would mean that an energy 32 times that of the Hanshin Earthquake would be attacking those nuclear reactors. And Professor Koyama Masato of Shizuoka University warns that while the Hanshin Earthquake continued for a little more than 10 seconds, the Tokai Earthquake might continue for as long as 1 or 2 minutes. If an earthquake that can topple superhighways in ten seconds continues for two minutes, what would happen?

The critical question is, if an earthquake on that scale were to occur, could the Hamaoka Plant safely shut down its reactors?

As soon as an earthquake occurs at the site of a nuclear plant, control rods must be inserted between the fuel rods to achieve emergency automatic shutdown. This is to stop the fission that was taking place in the uranium fuel.

Fukushima Daiichi Nuclear Plant went through emergency automatic shutdown. And it seems that the reactors themselves withstood the shaking of the earthquake. However according to the analysis of Tanaka Mitsuhiko which I reported above, and which I believe is irrefutable, this relatively mild (at the site of the power plant) tremor of around 500 Gals caused a fatal breakdown in the cooling system, something that should absolutely never be allowed to happen. This means that if a major earthquake occurs directly beneath the Hamaoka Plant, we can be 100% sure its plumbing system will suffer serious damage, which will very quickly lead to the melting of the fuel and its descent through the bottom of the reactor: meltdown.

Putting the reactor through emergency shutdown and stopping the fission in the uranium would not prevent this. The decay heat produced by the nuclear fuel is sufficient to melt the fuel rods; the incandescent fuel then melts through the bottom of the pressure vessel, comes into contact with water, and produces a hydrogen explosion. That is, within a few hours a greater catastrophe than happened at Fukushima would occur.

There is another scenario that can be thought of, which would make it a different type of accident from what happened at Fukushima. It is a description of what would happen if the emergency automatic shutdown system failed, and the fission in the fuel could not be stopped. With this, the reactor goes into a critical state. Critical state means that nuclear fission chain reaction continues. If this gets out of control and reaches the point of explosion, this means that the power plant becomes a nuclear bomb. That is what happened at Chernobyl.

How could the emergency shutdown system fail? Both the Fukushima and the Hamaoka reactors are the boiling water type (BWR). In this type of reactor the steam used to operate the turbine is taken from the top of the pressure vessel, which is why the control rods are inserted from the bottom. (See Diagram #2) The control rods, though they are called rods, are not simply like square pieces of lumber, but have a special shape with many blades attached to them like many crosses. These are inserted into narrow crevices between the fuel rods.

If the earthquake struck at the expected magnitude exceeding 8, the staff in the control room would not be able to remain standing. This level of earthquake makes people feel as though they are floating in the air. In the case of the Tokai Earthquake, there will be both vertical and lateral tremors simultaneously. Under this complicated shaking, the equipment for inserting the control rods cannot be expected to work properly. Undergoing vertical tremors greater than the BWR reactor was built for, the whole bundle of fuel rods could easily be bounced up into the air.

There is nothing built in to prevent that. And while the lateral tremors are continuing, the control rods won't go in. Moreover, both the fuel and the control rods could easily get bent out of shape and jam together. Even if a control rod is successfully inserted, continuing tremors may cause it to fall out again.

In fact, there have in the past been cases in Japan where control rods have dropped out of BWRs and caused accidents, but this was kept secret until March, 2007. On 2 November, 1978, five control rods fell out of the #3 reactor at the Fukushima Plant, and put the reactor into a critical state that lasted for seven and a half hours. On 18 June, 1999, three control rods dropped out of the then shut down #3 reactor at the Shika Plant in Ishikawa Prefecture, also creating a critical condition. Because these accidents were so dangerous, the electric companies and the nuclear industry concealed them from the public. Judging from this only, we can understand that these are bands of dangerous criminals. After the meltdown accident at Fukushima, there were people who apologized, but this is not the sort of thing that can be settled with an apology. They are a dank and moldy tribe, who never think of the safety of the public. These people, for whom the fact that so long as the nuclear power industry exists they can make money from it is more important than the lives of their countrymen, have jobs as professors in universities or as nuclear/radiology specialists, and are lying to this day. In other words, all the 54 nuclear plants in Japan are defective equipment, because we know that a big earthquake will easily bring any of them to meltdown.

Moreover, in May, 1991, three control rods dropped out of Hamaoka's #3 reactor while it was shut down for a periodic inspection. And in December, 2000, again three rods fell out of Hamaoka's #1 reactor. When these accidents happened, there was no earthquake.

If an earthquake is so big as to make emergency shutdown impossible, the pipes for the cooling system, which are more fragile than the reactor itself, will be damaged first. Water and steam will gush out explosively, and it will become impossible to cool the reactor. At Fukushima, after the fission in the nuclear fuel was stopped, the cooling system was lost, and the decay heat could not be removed, leading to one crisis after another. If the Tokai Earthquake hits Hamaoka and both the emergency shutdown system and the cooling system fail, then merely thinking about what kind of explosion would follow makes one turn cold. When I say the worst will happen, I will not add the evasive word "possibly"; I mean, without fail.

Safety Second; Proceed With the Plan

Why was the Hamaoka built at such a dangerous place? It is closely related to the timing of the acceptance of 'plate tectonics theory' (explained in Chapter 4) in Japan.

It was in 1959 when the Electric Power Development Coordination Council (attached to the Prime Minister's Office, since abolished)approved the construction, at Tokai Village, of Japan's very first industrial nuclear plant. As I mentioned above, plate tectonics theory was established by Tuzo Wilson, a Canadian scholar, in 1968. That is, theoretical proof that major earthquakes occur regularly around the periphery of the Pacific Ocean appeared nine years after the decision to build a nuclear plant at Tokai Village.

Just-completed plate tectonics theory was the cutting edge of science; Japan did not, on the basis of it, reconsider its nuclear power program, which was already under way. Part of the reason was that the Japanese archipelago is, geologically, relatively new, and it was difficult to prove that the continental theory of plate tectonics applied to the Japanese case. Also there was the feeling among Japanese scholars that as Japan has so many earthquakes, Japanese earthquake science must be the most advanced, which produced an attitude of, What do these Western scientists know about earthquakes!?

In this atmosphere, the construction of the Hamaoka Nuclear Plant was decided in 1969 and begun in 1971. The following year Prof. Sugimura Arata, formerly of Kobe University, published a paper "Plate Boundaries in the Vicinity of Japan" in which he demonstrated, using plate tectonics theory, that there is large scale movement of the earth's crust near the Izu Peninsula. However the nuclear power plant program was never reviewed. Needless to say, whenever they construct a nuclear plant, they do a geological survey of the site. But that takes place after the electric company has already chosen the site, as a kind of alibi to enable them to say that the site is safe.

Electric Companies don't choose a site because the ground is safe, but because the land is easy to get hold of and the local people are cooperative. And the companies give massive financial support to the academic associations to which the geologists and seismologists who do these surveys belong. Many of these scholars are hired as advisors by the electric companies, at which point they sell their souls as scholars. So even if the result of a feasibility study shows danger, they will never say "stop".

Baseless Elevation of Earthquake Resistance Standards

In the 1980s, as plate tectonics became the common view among scientists around the world, the theory achieved mainstream recognition in Japan as well. It had been vaguely assumed that even if the Hamaoka Plant should be hit by an earthquake, its seismic center would be at Enshunada which is somewhat distant from the plant. From the 80s, however, seismologist Ishibashi Katsuhiko has been pointing out that its center would be the trough beneath Suruga Bay, which is right nearby, which means that a major plate-type earthquake would occur directly beneath the plant.

In 2004, Inamura Kazuo, the founder of Kyosera (Kyoto Ceramic Co.

Ltd.) became a supporter of the "Nationwide Petition to Prevent a Genpatsu Shinsai". Government Headquarters for Earthquake Research Promotion has made public its official view that there is an 87% probability that the Tokai Earthquake will happen some time in the next 30 years. These days, this government organization goes so far as to say that if the three impending major earthquakes, the Tokai, the Tonankai, and the Nankai should all occur at the same time, it would cause economic damage of 81 trillion yen and death toll of 25,000, and would threaten the existence of the nation. However when you consider that such a genpatsu shinsai syndrome would destroy both the Tokyo metropolitan area and the Chubu Nagoya economic zone, you must admit that the government prediction is far too mild.

The ability of the Hamaoka Plant to withstand earthquakes has become the subject of vigorous debate, and standards for its seismic adequacy have been gradually revised. Reactor #1 was built to withstand an earthquake of 450 Gal; the standard was raised to 600 Gal for reactor #3. Later the standard was raised to 800 Gal, and today the plants have been reinforced so as to withstand an earthquake shock of 1000 Gal. Doesn't this process seem rather strange?

Compared to the 70s when it started its operation, the Hamaoka plant is more than twice as earthquake resistant. This may be a matter for rejoicing, but if you think about it, what then was the meaning of the tale we heard all along, that nuclear power is absolutely safe? Put simply, it means that all those years they were lying. Is there any reason to believe that the same people are not lying now?

There is no basis for believing that at the current standard, 1,000 Gal, safety can be assured. The reason is that 1,000 Gal surpasses the force with which gravity accelerates falling objects, which is 980 Gal. If vertical tremors of 1,000 Gal force should occur, everything will float up into the air. The buildings housing the reactors will fly off their foundations.

What kind of reinforcement could be added to buildings that were built to withstand 450 or 600 Gal earthquakes, to enable them to survive something of that force? For something as complex as a nuclear power plant, adding reinforcements here and there has no meaning.

Some may think that there will be no earthquake of 1,000 Gal. But just three years ago, during the 2008 Iwate-Miyagi Inland Earthquake, vertical movement of 3,866 Gal was observed at Ichinoseki city. People who say that the Tokai Earthquake will bring no earth movement to the Hamaoka plant greater than 1000 Gal must have something wrong with their thinking. As for the operators at CHUDEN, they think of nothing at all except for the single fixed idea that the plants must be kept in operation. In reality, there is no nuclear power plant in the world that could withstand such an earthquake.

In fact, Hamaoka has already experienced earthquake damage that brought it to within a hairsbreadth of going critical.

This was during the 11 August, 2009 Suruga Bay Earthquake. All the reactors went into emergency shutdown mode. Cracks appeared in #5 reactor's turbine building, and a 15 square meter area along the outer wall sank as much as 10cm. Also, out of the approximately 250 control rods, the drive units for about 30 rods went out of order.

Reactor #5, which started operation in 2005, is a large-scale reactor of the latest design, and proudly advertised by CHUDEN as the most earthquake-resistant reactor in Japan. In this earthquake it experienced extraordinarily large shocks. East-west movement on the first floor was recorded at 488 Gal, which surpassed the S1 figure of 484 Gal (S1 is the figure, calculated in Gal, for "the strongest earthquake movement that can reasonably be expected" at each specific location). Also, on the third floor, east-west motion of 548 Gal was recorded which was close to the S1 figure of 625 Gal for that location. (Figure 15 on page 108)

This movement surpassed that experienced by reactors #3 and #4, which

Figure 15
Acceleration of movement at the Hamaoka Plant caused by Suruga Bay earthquake

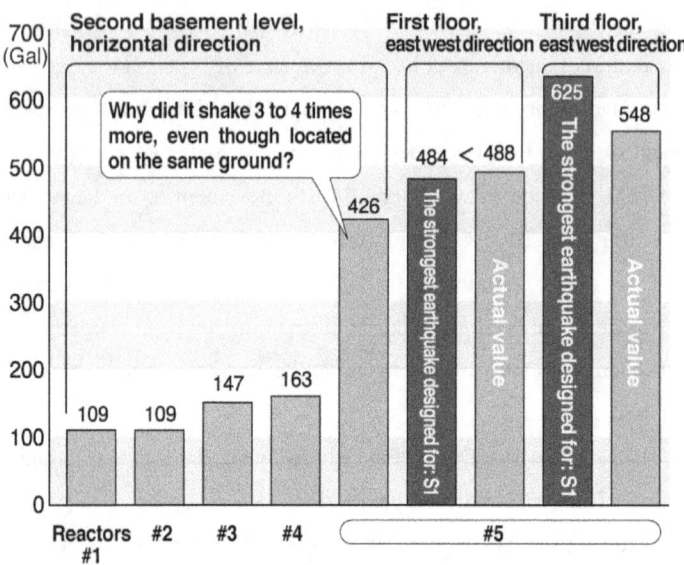

are located at the same site, by a wide margin. A serious question was raised both by local people and by scientists: Why would this be so in the reactor that is supposed to be state-of-the-art and the most earthquake resistant? Especially considering that the Suruga Bay Earthquake, which caused this movement, was relatively minor, with a magnitude of 6.5, only 126th that of the Great Kanto Earthquake.

Hardly anybody remembers the Suruga Bay Earthquake today. Other than Hamaoka Plant, there was little damage. It is sometimes called the Shizuoka Earthquake, because the Meteorological Bureau didn't give it an official name – it was that small. But this small earthquake caused movement that went beyond the design limits of the reactor. If the Tokai Earthquake were to occur beyond magnitude 8, the movement would be 178 to 1,000 times stronger than that of the Suruga Bay Earthquake. If it came with the force of the last Tokai Earthquake, Ansei-Tokai

Earthquake of 1854, energy 700 times stronger than that of the Suruga Bay Earthquake would directly hit the area. Since the Hamaoka Plant, with the "strongest" earthquake durability of all reactors in the nation, was damaged by such a small earthquake, we can assume that all the others would be destroyed entirely.

There is a further danger. The S1 figure, indicating "the largest earthquake force reasonably to be expected", is used as the standard of durability against stress for all of the machine parts used in nuclear power plants. This means that if any parts are subjected to stress exceeding that amount, even if no damage can be discovered by an external inspection, within the metal they may suffer invisible damage or deformation that cannot be repaired. Having worked as a technician to produce machine parts, I am fully aware of the risks of that kind of damage. If such parts continue to be used, major accidents could happen without an earthquake.

Today the people are entrusting their lives to the unreliable CHUDEN, and waiting for the Great Tokai Earthquake, which is bound to come.

Can a Sand Dune Protect You Against a Tsunami?

The Fukushima Daiichi accident demonstrated that even if the reactors succeed in going into emergency shutdown mode, if the cooling system is damaged a genpatsu shinsai syndrome can occur. As I explained in Chapter 1, the cause the failure of the cooling system at Fukushima was the electrical failure brought about by the tsunami.

The Hamaoka Plant is also located on the coast, and as the Tokai Earthquake's seismic center will be under the sea, they need to be prepared for a possible Tsunami.

CHUDEN explains that "there is no risk of tsunami since there is a 10m high sand dune between the plant and the ocean." This is like a bad joke. In the 1854 Ansei-Tokai earthquake, the record shows that the tsunami went over the sand dune and swept over 600 meters inland. In the recent Great East Japan Earthquake, we were horrified by the image unfolding in front of us of the mass of sea water rushing across the Tohoku inland area, stopped by nothing and eating up everything in its path. The Hamaoka sand dune is the same as the sand piles built by children playing at the beach. It is no more than a breakwater built long ago along the Hamaoka coast by piling up sand. CHUDEN simply used bulldozers to raise the height to ten meters. In the years since then much of the sand has eroded away, and it is pretty broken down. Before a tsunami, which would easily flood over a concrete breakwater, this sand dune would be washed away in an instant, and nothing would remain.

Following the Fukushima meltdown incident, CHUDEN hurriedly announced that they were going to build a 12+ meter breakwater fence on the landward side of the sand dune. A lot of people who heard this mistakenly believed that they were going to build a strong 12 meter breakwater. That is not the case. It turned out that they are planning to build 12 meter fence inside the sand dune. Being criticized for this, they announced that they would make it 15 meters. A fence, just like a fence around a house . . . and they are not joking! When I learned that this was a serious countermeasure being proposed by an electric power company, I couldn't get my gaping mouth to close. The people who believe this would be a countermeasure against a tsunami operate the largest nuclear reactor in Japan, built on top of the most dangerous earthquake zone in the country. This sums up the earthquake preparedness of the Hamaoka Nuclear Power Plant.

Tsunamis are said to increase their heights at deeply indented rias coastlines (a rias coast is one, like much of Japan's, which is shaped by former river valleys that have sunk into the sea, forming deep harbors and estuaries). In the1983 Nihonkai Chubu Earthquake, I watched the tragic image on TV where children were washed away when the tsunami of more than 10 m in height hit the rectilinear coast of Yamagata, Akita, and Aomori. The coast near the Fukushima Daiichi Plant is not a rea shoreline either.

The Hamaoka Plant is designed to take in sea water to cool its reactors from a water intake tunnel dug under the sand dune. If a tsunami hit, this tunnel would be quickly clogged up with mud and sand. The tsunami flood would be followed by a powerful outgoing tide, which would be repeated several times. During the receding tide, even if the tunnel is not clogged, it would not be possible to take in sea water. In a nuclear power plant, there is a pool of sea water for cooling in times of emergency. However, there is only enough sea water in the pool to maintain the cooling function for 20 minutes.

It has been confirmed that beneath the site of the Hamaoka Plant there are multiple earthquake faults. If an earthquake were to occur at the same level as the1854 Ansei-Tokai earthquake, the whole area of Omaezaki would rise by 1 to 2 meters with the rebounding of the Eurasian plate. In the Great Kanto Earthquake, the southern tip of the Boso Peninsula in Chiba Prefecture rose by 4 m. Under the influence of that much energy, it is easy to imagine that the numerous faults running under the Hamaoka site would start to move, causing cracks in the ground, damaging buildings, plumbing, electric power sources, and water intake tunnels.

The more you think about it the more you wonder, Why, then would anyone build a nuclear power plant on such a site? – and the more disillusioned you become. That is the Hamaoka Plant.

More than half of the Japanese people live in the huge metropolitan/economic area that stretches from Tokyo through Chubu to Kansai. If a genpatsu shinsai syndrome should occur at the center of this zone, the damage will be unimaginable. People directly hit by the earthquake would be showered with radioactivity. There would be no place to escape to. If the fission in the uranium cannot be stopped, and continues through to meltdown and the critical state, the very center of Japanese society would be annihilated.

A radioactive cloud, even with a mild breeze of 2m per second, will travel 500km in three days. Thus, within a week, not only central Japan but the entire archipelago from Hokkaido to Okinawa will be irradiated and, of course, Japan's neighboring countries as well. Inside Japan transportation and transport systems will be cut, there will be nowhere to escape to. The Fukushima Daiichi Disaster was terrible. But the Hamaoka Plant contains the possibility of a catastrophe many times, or many tens of times, worse.

The Hamaoka Plant should be shut down. There is no time to debate. A major earthquake is impending. All the 54 reactors throughout the nation are located on sites where earthquakes are possible, but the Hamaoka Plant, in terms of scale and the urgency of the risk, is the worst. There is simply no doubt that the Great Tokai Earthquake will come in the near future.

As I mentioned above, #1 and #2 reactors at Hamaoka have been taken out of operation. Reactors #3, #4, and #5, which are running, have a capacity 3,617,000 kW in total. However, over the past 8 years, the Hamaoka Plant has operated on the average at about 50% of capacity, mostly due to inspections and troubles. Thus, it has only been generating about 1,800,000 kW. CHUDEN is currently building the Joetsu Thermal Plant in Joetsu city, Niigata Prefecture, which is to start operation in July, 2012. This is a large scale, high performance power generator using the cleanest energy, natural gas. Its output will be 2,380,000 kW, which is enough to offset the loss of Hamaoka's nuclear generated power. If electricity is supplied by building up this kind of thermal plant, the Hamaoka Plant and the other nuclear plants will be completely unnecessary. And Japan would be liberated from the threat of the genpatsu shinsai syndrome.

If Toyota Motors, which is located in this area, does not take the lead in the "Shut Down Hamaoka" movement, all Toyota owners should take its business managers to task for being slow of wit. It makes no sense to go driving around in an irradiated area in your Prius. Most business managers in Japan need to learn how to do basic thinking. The future of the electric car, which is dependent on the all-night operation of nuclear power plants has, along with that of the all-electric home, entirely vanished. Song Masayoshi, the President of Softbank, has expressed his intention to oppose nuclear power. This is the most natural human response.

Chapter 6:

A Nuclear Power Archipelago over Active Earthquake Faults

As the Japanese Archipelago has entered a period of frequent earthquakes, a massive earthquake might occur anytime and anywhere, including the Tokai Earthquake and the Nankai Earthquake. I wrote above about the Hamaoka Nuclear Power Plant, the operation of which has to be stopped before any other nuclear plants in order not to reproduce the terrifying genpatsu shinsai syndrome.

As for the other nuclear power plants, then, are they safe?

At present, the Japanese Archipelago is crowded with fifty-four nuclear reactors. How many people know what location each reactor is built on and what safety measures have been taken? When looking carefully into these matters, one can come across surprising facts one after another.

They are all based on irresponsible geological surveys and underestimation of earthquake faults. The established Earthquake Resistance Standards are without scientific basis and therefore subject to arbitrary changes. Voices of anxious residents are noted neither by administrative organs nor judiciary authorities.

In this chapter, eighteen nuclear power plants are looked into. The expression "active fault", which is often used in this chapter, indicates a fault that has recurrently caused earthquakes over the past some hundred thousand years and is likely to cause quakes in the future.

Hokkaido: Tomari Nuclear Power Plant – Earthquake Fault Discovered Under the Sea, 70km Offshore

The Tomari Nuclear Power Station of Hokkaido Electric Power Co. is located on the coast of Tomari Village (population about 2,000) in the southwest of Hokkaido. Its #1 reactor began operation in 1989, and currently there are three reactors. Nowadays, reactors normally have a generating capacity ranging from 1.1 million to 1.3 million kW of electricity, whereas the #3 reactor of Tomari, which began operation on 22 December 2009, generates only 0.912 million kW. It is a pressurized water reactor built by Mitsubishi Heavy Industries, the type mainly used in western Japan, and is different from the boiling water reactors that are installed in the Fukushima and Hamaoka nuclear power plants. There hadn't been any orders for pressured water reactor for a long time and its manufacturing technology had almost disappeared, when the Tomari Power Station brought in that old-style reactor.

Many people, when hearing of southwest Hokkaido, may recall the Southwest Hokkaido Offshore Earthquake.. That 1993 earthquake with a magnitude of 7.8 is known as the largest on the Japan Sea side of the archipelago in recorded history.

The most damaged area was Okushiri Island. That island was hit by a disastrous tsunami more than 30m high, and out of a total population of 4000, the number of dead and missing reached 230 according to a report made 10 years later.

As indicated by such a major earthquake, the southwest of Hokkaido and the area offshore from it are an active earthquake zone. This zone extends from Sakhalin to Hokkaido's Mt. Usu. A survey of the topography of the ocean floor conducted between 1998 and 2001 by the National Institute of Advanced Industrial Science and Technology, using a submarine, revealed a "seismic gap" of some 50 km in length north of the epicenter

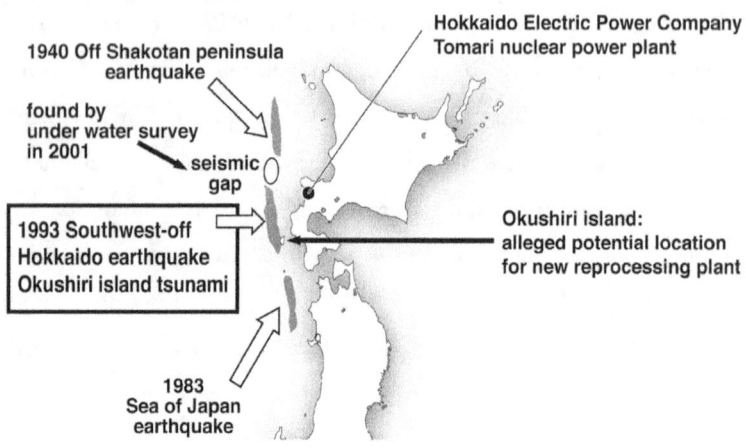

Figure 16
North part of Japan
Main focal regions on Japan Sea side

Hokkaido Electric Power Company
Tomari nuclear power plant

1940 Off Shakotan peninsula
earthquake

found by
under water survey
in 2001

seismic
gap

1993 Southwest-off
Hokkaido earthquake
Okushiri island tsunami

Okushiri island:
alleged potential location
for new reprocessing plant

1983
Sea of Japan
earthquake

of the Southwest Hokkaido Offshore Earthquake (figure 16). This is the area where seismic energy is presumably accumulating, so the probability of a large-scale earthquake occurring there is high.

Furthermore Watanabe Mitsuhisa, a professor of Toyo University and his group pointed out in 2009 that there lies another active fault of 60 to 70 km in length on the ocean floor 15 km west of the Tomari Nuclear Power Plant. Professor Watanabe is a specialist of change topography, capable of understanding the topography of seabed by observing rises and deformations of coasts and applying them to the historic trend of dynamic landform mechanism. For us, he seems to be a scholar gifted with clairvoyance. This fault is the closest to the Tomari Nuclear Power Plant among many other sea faults in the vicinity and were an earthquake to occur on it, it is likely to be of a magnitude of 7.5 or more.

In spite of being on such a precarious location, the Tomari Nuclear Power Plant was designed and built with Earthquake Resistance Standards of as low as 370 Gal. The standard was raised to 550 Gal in 2006, for fear of a possible earthquake directly underneath the site, in accordance with the New Guideline to Review Earthquake-Resistant Designs. This raised standard, however, is of course nowhere nearly sufficient to withstand a large-scale earthquake of a magnitude of 7.5. While the review as to whether the reinforcement works are sufficient is still going on, the plant continues in operation. It is like a car without a safety inspection running freely on public roads.

Aomori: Higashidori Nuclear Power Plant – Operating at the Country's Lowest Anti-earthquake Safety Level

The Higashidori Nuclear Power Plant of Tohoku Electric Power Co., located in Higashidori Village (population about 7,000) of Aomori Prefecture, started its #1 reactor in 2005 with a generating capacity of 1.11 million kW. At present, there is only one reactor in operation. On the neighboring plot, TEPCO started constructing its own #1 reactor in January 2011, but work has been suspended due to the meltdown accident of Fukushima Daiichi Nuclear Power Plant.

As is well known, on 7 April almost a month after the Great Tohoku Earthquake, the biggest aftershock following 3/11 took place and the whole area of four prefectures - Iwate, Aomori, Yamagata and Akita - had a power failure. At that time, at the #1 reactor of the Higasidori Nuclear Power Plant the emergency generator kicked in as the external power supply was cut off. The generator, however, broke down after the external power supply was restored, which means all three emergency generators including the two that were undergoing maintenance at the time, were out of order. The situation was at the verge of a second Fukushima.

The Higashidori Nuclear Power Plant is on the Pacific Coast side of Shimokita Peninsula, which is known to have been under the sea until the Jomon Era, a few thousand years ago, and the ground is unstable. In particular, beneath the site of the TEPCO plant run three faults; one of them, called the J-1 Fault, runs directly beneath the construction site, and is a large fault as much a 400m in depth. It is easy to predict that, in case of an earthquake, the earth there will crack, differences in level will appear, and the soil will liquefy. It is hard to understand the mentality of people who would choose such a site.

Nearby there is the Yokohama Fault, registered as an active fault. When it was pointed out that the fault is likely to cause a serious earthquake, an additional survey was required and the safety review was prolonged. In April 2010, however, NISA closed its first review, and NSC on 13 December, and the Nuclear Committee on the following 14 December, approved the granting of permission for the construction. It is chilling to recall that those who granted this permission are also the ones responsible for the Fukushima Meltdown. Furthermore, there is another 84km fault at the outer edge of the continental shelf, offshore from the Shimokita Peninsula. And just to the north of this offshore fault a number of 10 km faults have been found, which probably all belong to a single fault; if so, they would add up to a massive fault of 100km or more in length. So it would be no surprise if an earthquake of magnitude 8 or above, that is, on the scale of the Ansei Tokai Earthquake, were to strike.

Precarious as the location of the Higashidori Nuclear Power Plant is, the Earthquake Resistance Standard of the plant is set at only 450 Gal, still the lowest among all plants in Japan even after it was raised in accordance with the New Guideline to Review Earthquake-resistant Design (the standard had been set at 375 Gal before the New Guideline was issued in 2006).

Why is the Earthquake Resistance Standard of the plant kept as low as this? It is because the standard needs to be consistent with that of the Rokkasho Nuclear Fuel Reprocessing Facility which is located at Rokkasho Village on the same Shimokita Peninsula.

In Rokkasho Village (about which I will have more to say in the final chapter) the Rokkasho Nuclear Fuel Reprocessing Facility started test operations in 2006 for reprocessing spent nuclear fuels and its interior is now highly contaminated by radiation from plutonium and high-level radioactive waste. Because of this danger, access is limited, which makes it difficult to do reinforcement work there, as a result of which the Rokkasho Plant's Earthquake Resistance Standard remains at 450 Gal as before. Raising the standard at Higashidori plant would inevitably lead to questioning the standard of Rokkasho plant, as the two plants are located near each other. In order to insist that the neighboring Rokkasho plant is safe at 450 Gal, the Earthquake Resistance Standard of Higashidori plant also remains unchanged at the same low level.

As difficult as this is to believe, there is more. The Tohoku Electric Power Co. has announced that it would begin "extended cycle operation" from July 2011. All this means is that the maximum interval between periodic inspections is to be extended from 13 months to 16 months, so that longer operation without interruption will be possible. What it amounts to is organized negligence.

Aomori: Oma Nuclear Power Plant – Scheduled to Open in 2014 in a Volcano Zone

In Aomori Prefecture, there is another nuclear power plant under construction besides the Higashidori plant. Its location is Oma City (population about 6,000), which is a fishing village on the northern edge of Shimokita Peninsula, and well known for Oma tuna.

The construction of the Oma Nuclear Power Plant was promoted by the Electric Power Development Company (J-POWER), and construction began in May, 2008. Operations are scheduled to begin in November 2014.

Oma plant is to be the world's first commercial plant to use only MOX fuel – a mixed oxide of plutonium and uranium (see Chapter 2). A very large amount of plutonium will be on the plant site.

From the viewpoint of topography, its location is near the tip of Shimokita Peninsula, with the sea both to the west and to the northeast. On each side, active faults of 50 km and 45 km respectively, have been found under the sea by scholars of change topography: Prof. Watanabe Mitsuhisa of Toyo University, Prof. Nakata Takashi of Hiroshima Institute of Technology, and Prof. Suzuki Yasuhiro of Nagoya University. These faults are believed to be the traces of large-scale earthquakes of about magnitude 7 having taken place at least twice in the past 7,000 years.

Like the other areas of Shimokita Peninsula, Oma City was under the sea in the Jomon Era and its foundation is not solid. Also, the area is a volcanic zone with three volcanoes including Mt. Osore within 30 km on the Aomori side and two including Mt. E on the Hokkaido side. It is difficult to understand why they are building a nuclear power plant at such a location.

The Earthquake Resistance Standard of the Oma Nuclear Power Plant is, in spite of the situation described above, to be only 450 Gal which is, together with the other nuclear plants on Shimokita Peninsula, the lowest in Japan.

Aomori: Rokkasho Village Reprocessing Plant – More Dangerous Than Any Nuclear Power Plant

A little to the south of the Higashidori Nuclear Power Plant is Rokkasho Village (population about 11,000), where the Rokkasho Nuclear Fuel Reprocessing Plant is located, managed and operated by Japan Nuclear Fuel Limited. This facility is, in accordance with the Nuclear Fuel Cycle Program, a chemical plant designed to extract uranium and plutonium from spent nuclear fuel from all the nuclear power stations in Japan. The Rokkasho plant is a large factory full of many pipelines that stretch in all directions. The total length of the pipelines, through which radioactive material flows, reaches 1,300 km. (The pipelines for uranium and plutonium alone amount to 60 km).

What that means is that this plant is the most vulnerable facility to a disaster, such as an earthquake, where a powerful force is exerted. According to Japan Nuclear Fuel Limited, there are as many as 26,000 joints in these pipelines, and a rupture of any of those joints could lead to leakage of high-level radioactive material.

However, at this plant, where of all places safety should be given the greatest priority, the Earthquake Resistance Standard is set at only 450 Gal. This standard is remarkably low compared with those of most other nuclear plants in Japan where the standard was raised in 2006 to at least 600 Gal in accordance with the New Guideline to Review Earthquake-resistant Design.

The reason for the low standard is, as already mentioned, that the facility had already started active test operation in March 2006. Once the reprocessing facility started operation, there was very limited access to the interior because of high radiation levels, which makes reinforcement work extremely difficult. By adding what reinforcement they could from the outside, the standard was barely raised from the original 375 Gal to 450 Gal, but that is no more than just a calculation. In order to insist that "the safety of the reprocessing facility is ensured at that standard," the resistance standard of the nearby Higashidori Nuclear Power Plant is also kept at 450 Gal.

In addition to worries about its earthquake resistance, the location of Rokkasho reprocessing facility is also extremely unsafe. In 2008, an active fault of at least 15 km was newly discovered directly beneath the facility. It was again Watanabe Mitsuhisa of Toyo University, Nakadai Takashi of Hiroshima Institute of Technology and Suzuki Yasuhiro of Nagoya University who conducted the survey, the findings of which were widely carried in the newspapers. Also, it is certain that this fault is connected with the fault at the outer edge of the continental shelf which runs for 84

Figure 17
Shimokita Peninsula was under the ocean
and is a very vulnerable area

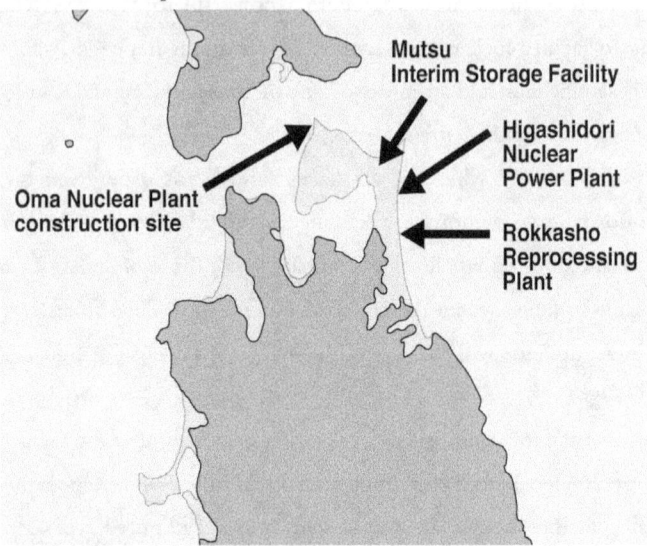

Mutsu
Interim Storage Facility

Higashidori
Nuclear
Power Plant

Oma Nuclear Plant
construction site

Rokkasho
Reprocessing
Plant

"Medemiru Nippon Retto no Oitachi Ochiri Zukan"
(Visual history of the Japanese Archipelago - a picture book of ancient geography)
Editorial Supervisor: Masao Minato
Tsukijishokan, 1978

km under the Pacific Ocean off the coast of Shimokita Peninsula. The combined length of these faults amounts to more than 100km from land to sea floor. This fault, therefore, is capable of causing a magnitude 8 earthquake underneath the region.

As with the Higashidori Plant, the location was under the sea in the Jomon Era and the foundation is soft with many small faults besides the major active faults. There are faults not only beneath the reprocessing facility but also beneath what is called the Glass Solidification Warehouse, where high-level radioactive waste is stored. Nowhere are more hazardous materials stored than here, and nowhere is there a location more dangerous. Japan Nuclear Fuel Limited explains that the location was selected based on a careful survey, but facts keep coming out one after another that make us wonder if they actually conducted a survey at all. (Figure 17)

In fact, when the Sanriku Haruka Offshore Earthquake occurred in 1994 with the magnitude of 7.6, the Obuchi fishing port near Rokkasho Village suffered heavy damage such as the complete destruction of the concrete quay walls even though the epicenter was 180 km away under the Pacific Ocean. What would happen if an earthquake were caused by a fault closer to the coast? The Rokkasho Nuclear Fuel Reprocessing Facility is not a nuclear power plant, but it contains the possibility of causing a more destructive genpatsu shinsai syndrome than any of the nuclear power plants.

I will get back to the Rokkasho Nuclear Fuel Reprocessing Plant in the last chapter.

Miyagi: Onagawa Nuclear Power Plant – Was It Just Luck that It Survived the Great Tohoku Earthquake?

The Onagawa Nuclear Power Plant of Tohoku Electric Power Co. is located on the Pacific coast of Miyagi Prefecture. The plant spans two cities: Onagawa City (population about 10 thousand) and Ishimaki City (population about 160 thousand)

It started operation in 1984, and at present has three reactors generating electricity. In the Great Tohoku Earthquake, in many places the tsunami inflicted more damage than did the quake. In Onagawa City, too, the sea water surged up at a height of 17.6m according to the most recent data.

Fortunately, the Onagawa Nuclear Power Plant was built on a hill 14.8m above the sea level, and it is reported that only a part of the facility was inundated and that the reactors safely went into automatic emergency shutdown mode. The damage inflicted both by the tremor and the tsunami, however, has not been made clear. Besides, when on 7 April, about a month after 3/11, the big aftershock shut down the power in Iwate, Aomori, Yamagata and Akita Prefectures, it hit the Onagawa Nuclear Power Plant with an intensity of 6 on the Japanese Seismic Scale and "two out of three" of the external power supplies were reportedly stopped. The reactors were cooled with the remaining one. As it happened, they successfully got to the other end of the tightrope. Onagawa came that close to becoming the next Fukushima.

The Onagawa Nuclear Power Plant has experienced many troubles from earthquakes in the past. In the first place, Onagawa is located in an earthquake-prone region, called "The Oga Peninsula - Ojika Peninsula Tectonic Zone". Around this area, there have occurred many earthquakes in recent years, including the 2008 Iwate-Miyagi Inland Earthquake.

As for troubles accompanying earthquakes, first there are those that followed the 1993 Northern Miyagi Prefecture Earthquake. Though it was a small-scale earthquake of magnitude 5.8, the #1 reactor had a sudden increase in power output and came to an emergency shutdown. This was a typical phenomenon for a boiling-water reactor, where an earthquake causes an abrupt change of the flow and density of the water surrounding the nuclear fuel rods which sometimes brings about the sudden rise of power output due to the increase of neutrons that prompt the fission reaction. The same trouble has occurred in the Fukushima plant, where boiling-water reactors are also used.

What we can learn from this is that even in the case of a minor earthquake, which does not damage the reactors or rupture the pipelines, a reactor can go out of control. A nuclear reactor is a delicate piece of equipment. The worst accident brought about by a reactor going out of control – in which the reactor went critical – was the Chernobyl accident. Adding reinforcements to the outside of its container do not prevent this. Also, with regard to earthquake resistance, Onagawa Nuclear Power Plant is very precarious. During the 2003 Sanriku South Earthquake with a magnitude of 7.1, the plant recorded tremors exceeding the Extreme Design Basis Earthquake S2 (S1 is "the strongest earthquake reasonably to be expected" at a given site; S2 is "the strongest earthquake that can be imagined" at that site). When the 2005 Miyagi Offshore Earthquake of a magnitude of 7.2 occurred with the epicenter in the Oga Peninsula - Ojika Peninsula Tectonic Zone, the structure for housing the #1 reactor recorded a vibrational acceleration beyond the maximum that had been set assuming a magnitude of 7.5. There, 888 Gal was recorded, which far exceeded the 683 Gal S2 figure that had been established "just in case though in reality there is little possibility of it occurring."

Because of these events, the standard for vibrational acceleration at Onagawa Nuclear Power Plant was raised 1.5 times from 375 to 580 Gal, that is, by 1.5 times in accordance with the 2006 revision of the Guideline to Review Earthquake-Resistant Design. The officials had asserted that "the plant was definitely safe at 375 Gal", but this revision proved, after all, that their assertion had been groundless. Furthermore, the 2005 Miyagi Offshore Earthquake showed that the raising of the Earthquake Resistant Standard didn't necessarily guarantee the safety of the plant.

Anyway, the issue can not be settled by saying, how lucky it was that the plant wasn't damaged by the 3/11 earthquake. The plant remains in the middle of a zone of frequent earthquakes, and it has not been shut down.

Fukushima Daini Nuclear Power Plant – Located to the South of the Fukushima Daiichi Nuclear Power Plant and Next to the Large Futaba Fault

TEPCO's Fukushima Daini Nuclear Power Plant is only about 12km south of the Fukushima Daiichi Plant that caused the recent genpatsu shinsai syndrome. It spans two cities: Tomioka City (population about 16,000) and Naraha City (population about 7,500) of Futaba County in Fukushima Prefecture. (These numbers are the populations before the

Great Tohoku Earthquake)

The #1 reactor of Fukushima Daini Nuclear Power Plant started operation in 1982, and currently there are four reactors working. During the Great Tohoku Earthquake, this plant also incurred serious damage from the tsunami and lost its cooling function. At present, these reactors are reportedly in cold shutdown.

All the reactors of the Fukushima Daiichi Plant are sure to be decommissioned, whereas the Fukushima Daini Plant, somehow, is likely to resume operation after some restoration work. Few anti-earthquake measures, however, have been taken at this plant.

The Fukushima Plant is threatened not only by the Plate Boundary Earthquake which occurs in the Pacific offshore, as happened this time. On the inland side of the plant there is a large, active fault more than 70km in length, called Futaba Fault. This fault extends as far as southern Miyagi Prefecture. TEPCO, underestimating the length of the fault, has insisted the plant is safe. That is untrue. Massive aftershocks have been frequently striking this region after 3/11, and many people are concerned about the repair work going on at the Fukushima Daiichi Plant. But every time these aftershocks happen in Fukushima, I worry rather that the Futaba Fault may shift, and the Futaba Fault is closer to Fukushima Daini than to Fukushima Daiichi. It is easy to understand that this fault, which runs right in front of Fukushima Daini, will be affected by the movement of the Pacific Plate. Should this fault shift, a large-scale earthquake could happen directly underneath the inland area with a magnitude of about 7.9 (as large as the Great Kanto Earthquake). In that case, a more severe shock than the Great Tohoku Earthquake, which happened offshore, would strike the Fukushima Daini Nuclear Power Plant which, having been built without taking this fault seriously, would be blown away.

Compared with the facilities of Fukushima Daiichi that are worn out after 40 years of operation, those of Fukushima Daini are still new, but its four reactors have a large generating capacity of 1.1 million kW, so it would become a much larger disaster should radioactivity be released. Furthermore, in 2012 the #1 reactor of Fukushima Daini will have completed 30 years operation, which can be considered, for a machine, a lifespan. For it is common sense in the manufacturing world that no metal materials can be guaranteed for longer than 30 years.

In 2000, there was an accident at the #6 reactor of Fukushima Daiichi Plant in which a pipe was ruptured by a tremor of *shindo* 4 (on the Japanese seismic scale) caused by a minor earthquake offshore from Ibaraki Prefecture. The pipeline, however, had supposedly been designed to withstand an earthquake of *shindo* 6. Over time, the reactors of Fukushima Daiichi must have deteriorated. Likewise, the reactors at Fukushima Daini must also have deteriorated over time. Many people today don't remember this, but on 6 January 1989 the recirculation pump for the reactor core coolant for the #3 reactor of Fukushima Daini was badly damaged, bringing the reactor to the verge of meltdown.

Ibaraki: Tokai Daini Nuclear Power Plant – Inexcusable Deception of "Safety Assessment"

Tokai Daini Nuclear Power Plant of Japan Atomic Power Co. (JAPC) is located on the Pacific coast of Ibaraki Prefecture. It is in Tokai Village (population 37,000) and is the nearest nuclear power plant to Tokyo.

The plant has "Daini" ("number two") in its formal name because there used to be the Tokai Nuclear Power Plant, also of JAPC, on a neighboring site, the first commercial nuclear power plant in Japan. Tokai Village, therefore, is the birthplace of nuclear power generation in Japan. The first nuclear reactor was shut down 1998 and currently is being dismantled. The second reactor began operations in 1978 and therefore is a relatively old plant.

Ibaraki Prefecture in which the plant is located has been called "the nest of earthquakes". The region was long under the sea and is characterized by lakes and marshes; the ground is soft and unstable.

Inland from the coast where the plant is located, faults have been identified: the Kanto Plain Northwest Border Fault Zone, capable of bringing about an earthquake of magnitude 8, and the Sekiya Fault, capable of producing one of magnitude 7.5. If an earthquake were to occur on these faults, it would be directly underneath the area.

JAPC, however, hasn't properly evaluated the danger of these faults. In the interim report on seismic safety it issued in 2008, the large faults are divided and evaluated as many small, individual faults.

As a predicted magnitude of an earthquake is calculated based on the length of nearby faults, estimating the faults' length as short can produce the conclusion that a major earthquake will not take place. There are countless cases however– one of them being the Southern Hyogo Prefecture Earthquake - in which faults that seemed from the earth's surface to be separate, moved simultaneously as one fault when an earthquake came. In safety assessments for planned construction sites of not only the Tokai Plant but also all the other plants as well, when safety evaluations have been made of the planned sites, faults have routinely been cut up into short pieces, thus producing the result than only small earthquakes are to be expected.

With such assessments as these, the safety of nuclear power plants cannot be guaranteed. This unreal safety theory held up during the quiet earthquake period that continued until recently, but the cover-up has been unveiled now that an active earthquake period has come. One after another, cases are occurring in which nuclear power plants incur damage every time an earthquake comes. A typical example is the Kashiwazaki-Kariwa Nuclear Power Plant, which I will explain in the next section.

Niigata, Kashiwazaki-Kariwa Nuclear Power Plant – Incurred Critical Damage "Beyond Prediction" from the Chuetsu Offshore Earthquake

It was in 1985 when the #1 reactor at TEPCO's Kashiwazaki-Kariwa Nuclear Power Plant started running. There are seven reactors there at present and their generating capacity, aggregated, makes the plant the largest nuclear power plant in the world. Its location spans Kashiwazaki City (population about 91,000) and Kariwa City (population about 5000) of Niigata Prefecture.

The Kashiwazaki-Kariwa Plant has often experienced suspensions of operation due to earthquakes and accidents. In particular, when the 2007 Chuetsu Offshore Earthquake occurred, transformers caught fire and caused serious damage. It took two years before the plant could resume operation and even now not all the reactors are working; four reactors are running as of April 2011. But these four reactors have been put in operation despite warnings from many scientists and engineers that they are perilous and their operation should not be resumed.

Who took the initiative to resume operation? It was Haruki Madarame, a professor of the Graduate School of Tokyo University, who took the position of the chair of the "Chuetsu Offshore Earthquake Nuclear Installations Investigation and Study Sub-committee" under the Ministry of Economy, Trade and Industry. He is also the chair of NSC, which has been blamed for the Fukushima meltdown. During the injunction litigation of Hamaoka Nuclear Power Plant through which citizens were seeking to get the plant shut down, he took the witness stand for the defendant, Chubu Electric Power Co. and said repeatedly, as a good public relations man, "Nuclear power plants are safe". In Hamaoka, he is called Haruki Detarame ("detarame" meaning "haphazard" or "nonsense" in Japanese)". He took the initiative in resuming the Kashiwazaki Kariwa Plant's operation by disregarding deformations in of the pressurized vessels caused by the Chuetsu Offshore Earthquake.

Looking at the pictures of the Kashiwazaki-Kariwa Plant after the Chuetsu Offshore Earthquake, which show the building foundations exposed by violent cave-ins and fissures in the ground, and the joint of a huge crane truck broken, I felt it was lucky that the plant had barely survived. It means the earthquake shock that struck the plant was severe. In fact, the tremor at the building site measured at *shindo* 7, which is as high as the Japanese seismic scale goes, and workers in the control room had difficulty to keep standing during the quake.

The Chuetsu Offshore Earthquake, however, was just a middle-scale earthquake that occurred offshore with a magnitude of 6.8. The highest measurement of its force outside the plant was *shindo* 6 in Nagaoka City (remember the Japanese *shindo* scale measures the force of an earthquake at each specific location).. Why was the nuclear power plant alone stricken by a much greater tremor?

It is due to the location of the Kashiwazaki Kariwa Plant. Niigata was well known as one of a few places that produced oil in Japan. Kashiwazaki is at the center of that area and Nippon Oil Co. (now JX Nippon Oil & Energy Corporation) developed there. The ground of an area that produces oil is often composed of "soft rock", which will be broken into pieces under a little pressure. The Nishiyama Formation on which the Kashiwazaki Kariwa Plant is built is made of this very soft rock.

About half a year after the Chuetsu Offshore Earthquake, Prof. Watanabe Mitsuhisa of Toyo University and his group analyzed the graphic data, taken from a satellite by the Geographical Survey Institute, of the disaster-stricken area. From this they discovered that an area extending from the Madogasaka Fault, which is near the plant, all the way to Kashiwazaki City, and about 2km in width, had been raised some 10cm by the earthquake. That is, this soft foundation, moved by the energy of the earthquake, raised up the entire nuclear power plant site. That was why the unusual tremor of *shindo* 7 was recorded there.

That Kashiwazaki-Kariwa had such geological features was known before the plant was built. The oil refining in this area started in 1852 (Kanei 5 in the Edo Era) and since then, its geographical and geological features have been thoroughly examined in pursuit of oil production. Therefore, it is well known not only that the ground is soft but also that nearby there is an active fault, the Madokazaka Fault, and also that there are many small faults running under the site. The Nagaoka Plain Western Border Fault Zone, which runs north from Kashiwazaki City is as long as 92 km and capable of bringing about a massive earthquake with a magnitude of 8.1.

TEPCO, however, regarded the Madokazaka fault as already a "dead fault", and also completely disregarded the Nagaoka Plain Western Border Fault Zone, evaluating only the 17.5km Kehinomiya Fault. Afterwards, there were found another four faults under the sea, but continuation of the plant construction was made possible by underestimating the lengths of those as 10 km or less. That the plant, shaken in the Chuetsu Offshore Earthquake by tremors "beyond all prediction", came to the brink of genpatsu shinsai syndrome is the natural result of these decisions.

Following the Chuetsu Offshore Earthquake, the Earthquake Resistance Standard for Kashiwazaki-Kariwa Nuclear Power Plant was raised from 450 Gal to 2,300 Gal for #1 through #4 reactors, and to 1,209 Gal for #5 through #7 reactors – that is, by at most a factor of more than 5. Considering that the standard for other plants are about 600 Gal, these numbers are remarkably high. In fact, the tremor was measured at 606 Gal at the #2 reactor, 3.6 times as large as what had been assumed, and 2,058 Gal at the first floor of the structure housing #3 turbine. The Earthquake Resistance Standard could not remain below what had been actually measured.

Even though the Earthquake Resistance Standard of the plant has been raised and reinforcement work has been done the plant is still, as I wrote above, far from safe. Raising the resistance standard is just juggling the figures, and from an engineering standpoint, it is not realistic to expect that doing some reinforcing work is going to raise the actual earthquake resistance that much. Furthermore, when the 2004 Chuetsu Earthquake (an inland earthquake) occurred, the tremor measured a stupendous 2,515 Gal at Kawaguchi City in Niigata Prefecture. There have been found, also, new faults under the sea one after another that TEPCO has not included in its safety evaluation. Among those is the Sado Ocean Basin Eastern Border Fault, which is capable of causing an earthquake of magnitude 7.5. If an earthquake on that scale were to strike, seismic energy six times what is assumed by TEPCO could strike the plant. Even with the new, raised Earthquake Resistance Standard, the plant could not withstand that.

Another reason why the situation can't be considered safe is that the plant keeps using facilities that were subjected to the tremor of the Chuetsu Offshore Earthquake without undertaking proper inspection. Metallic parts that that have been subjected to force three times what they were designed for are likely to have internal deformations invisible from the outside. TEPCO, however, made fiberscope inspections on only two of the recirculation pumps, judged from that that all ten were undamaged, and resumed operations.

This resumption of operations should never have been allowed. I am anxious about the plant not only because of earthquake danger, but also because even without an earthquake a serious accident could occur any time given the current state of the Kashiwazaki-Kariwa Nuclear Power Plant.

Ishikawa/Shika Nuclear Power Plant - Kanazawa District Court Issued a Restraining Order Against Operation

The Shika Nuclear Power Plant of Hokuriku Electric Power Company (RIKUDEN) is located in Shika-machi, Ishikawa Prefecture (population about 22,000) on Noto Peninsula on the coast of the Japan Sea. Operation of its #1 reactor started in 1993. Currently, two reactors are in operation.

Noto Peninsula now sticks out from the main island of Honshu, but it is known that in the warm interval of the Jomon Era, the ocean level rose, and the area where the Shika plants are located were at the bottom of the sea.

Therefore, the ground is not firm and a proper geological survey would not produce data showing that it is a location where a nuclear plant can be built. Thus, when the surveyors bored into the ground and took out a core sample that showed the weakness of the ground, they would replace it with another that was firmer, thus fabricating the results. Normally, all core samples are marked with a serial number and stored. An in-house whistle blower, a person connected with the construction party, produced photographic evidence that a large number of core samples had been abandoned at the construction site. This became a major issue at the time. However, RIKUDEN refused all interviews and pushed through the plant construction.

The Shika Plant, built on an infirm ground, will doubtless be seriously damaged should a major earthquake should hit the area. The 2007 Noto Peninsula Earthquake was a medium size earthquake of magnitude 6.9. At that time, 711 Gal was measured at #1 reactor which was far beyond its aseismic standard of 600 Gal. It was nothing but a good luck that there was no major accident.

The Noto Peninsula Earthquake occurred on a seabed fault of about 20km long. RIKUDEN, however, underestimated this fault by dividing it into three faults of about 7km each. Also, despite the fact that the length of Ouchigata fault, which is very close to the plant, has been estimated to be about 44km by the government seismic research headquarters, RIKUDEN estimated its length as 8km. Later, in 2009, they changed their story, and announced that "it consists of two faults of 34 km and 20km." They are people who completely overlook the danger of earthquakes, and are not to be believed.

In 2006, Kanazawa District Court issued a restraining order against operation of reactor #2, which had just started up, which reads in part, "the fact that an active fault has not been confirmed does not mean that there can be no earthquake." This was an epoch-making court decision, where a claim brought by the general public held up at the district court level.

In 2009, however, the Kanazawa Branch of the Nagoya High Court overturned the 2006 decision of the Kanazawa District Court saying "The new aseismic code (made by the government after the lower court's ruling) reflects the latest perception. Its safety measures comply with the government code and no concrete risk is acknowledged." As in the example of the Shika Plant, so far in trials seeking to stop operation of nuclear power plants, judges one after another have served as government puppets, abandoning legal judgment grounded on law, and dismissing the people's claims with the words, "it complies with the government code". These judges, by declining to interfere with, or positively supporting, the government's nuclear policy, have exposed the lives of the people to great risk, and bear heavy responsibility for this. However following the Fukushima Daiichi Plant accident, we can expect that the "hands-off policy" of the judiciary will be subjected to severe scrutiny by the general public.

Wakasa: Genpatsu Ginza - On a Foundation of Multiple Faults, 14 Reactors All In Row

The coastline of Wakasa Gulf, Fukui Prefecture (Tsuruga, Mihama, Ooimachi, and Takahama-cho) is called Genpatsu Ginza [Translator's note: Nuke-Plant Ginza – Ginza being Tokyo's most famous and most glittering shopping street]. Along it stand five nuclear power plants, including the fast-breeding reactor (FBR) Monju, for a total of 14 reactors, a veritable ReactorLand. There is no other place in the country where so many plants are concentrated in one area. The following is a summary of each plant.

Japan Atomic Power Company (JAPC) / Tsuruga Plant

This plant is located on the east side of Tsuruga Peninsula, at the exit of Tsuruga Gulf, close to the cape. Number 1 reactor started operation on the opening day of the 1970 Osaka Expo. There are two reactors in this plant; #2 reactor has been in operation since 1987. Number 1 reactor is BWR and #2 is PWR. Currently, reactor #3 and #4 are under contemplation. Due to prolonged seismic safety inspection, there have been many postponements, and construction has not begun.

Kansai Electric Power Company (KEPCO) / Mihama Plant

Like the JAPC /Tsuruga Plant, this is located on Tsuruga Peninsula, only on the west side. This is an old plant; its #1 reactor started operation in the same year as the Tsuruga plant, 1970. Currently, three reactors are running. This is a typical aged plant, all of whose reactors started operation in the 70s.

Kansai Electric Power Company (KEPCO) / Takahama Plant

Of all the nuclear power plants around Wakasa Gulf, this is the farthest west. Its #1 reactor began operation in 1974. At present four reactors are in operation, but all are aging, dating back to before the mid-1980s.

Kansai Electric Power Company (KEPCO) / Oi Plant

This plant is located at the tip of Oshima Peninsula, which surrounds Kohama Gulf (it used to be an island). Operation of reactors #1 and

#2 started in 1979. On the other hand, reactors #3 and #4 started to run in the early 90s. This is a plant where old type and relatively new type reactors are mixed together. Among the nuclear plants located on the region to the west of Hokuriku tunnel, so-called "Southern Fukui", its power generation capacity is the greatest.

FBR Monju

This Fast-Breeding Reactor (FBR) is operated by the Japan Atomic Energy Agency (JAEA). It is a special reactor designed to reprocess spent nuclear fuel used by other power plants, and to use the plutonium so retrieved to generate electricity. This plant is located close to the tip of Tsuruga Peninsula facing Wakasa gulf, just between JAPC's Tsuruga plant and KEPCO's Mihama plant.

On the coast of Wakasa Gulf, ever since the 1948 Fukui earthquake (magnitude 7.1), no large scale earthquake has occurred. Up to today, there has been a seismic quiescence. Thus, no big accident has been caused by earthquakes here.

However, this region is not free from major earthquake risk. Rather, looking back into the prewar periods, prior to the Tonankai Earthquake (1944) and the Nankai Earthquake (1946) , the region has proven to be an area which has experienced major earthquakes about once every 40 years, ranging from magnitude 6.8 to 7.3 The 1927 Kita Tango Earthquake (magnitude 7.3) took a toll of 300 dead and missing. This situation resembles the series of mid-level earthquakes that have been happening in the Kanto and Tokaido areas, presaging the Tokai earthquake.

Looked at from that that point of view, big earthquakes of magnitude 7 level, including the 1995 Great Hanshin Earthquake and the 2000 West Tottori Earthquake have been striking from the Kinki (Osaka/Kobe) region up to the coast of the Japan Sea [Tottori is to the southwest of

Genpatsu Ginza, on the Japan Sea]. More than 60 years have passed since the last earthquake, the Tonankai Earthquake. Western Japan should also be seen as entering into an era of major shifts in the earth's crust. If so, then people should be prepared for big earthquakes coming one after another in the peripheral area of Wakasa Gulf, just like revival of the Fukui earthquake.

Geographically speaking, from the east bank of Lake Biwa across to Wakasa Gulf, there are numerous fault complexes. Above all, the Yanagase-Sekigahara fault belt is a huge active fault 100km in length. If an earthquake should occur on it, it would be a major one of the magnitude 8 class. Also, under the peripheral area of Tsuruga Gulf where Monju is built, active faults, including some under the sea, are particularly concentrated. Any of these has the capability of causing an earthquake at the magnitude 7.2 level.

It has been learned that Monju does not have fault "nearby"; rather it has a fault that cuts directly beneath the plant. This is the Shiraki / Nyu fault that the High Court judged "does not exist, when in 2005 it dismissed a suit brought by local residents. In 2008 however, three years after the ruling, JAEA nonchalantly admitted that it was there after all.

The length of this fault is 15 km, which means that it is capable of causing an earthquake of magnitude 6.8. Looked at from the ground level, it seems to run 500m to one side of Monju, but as it cuts the earth at a 60 degree angle, actually it runs right under Monju, 850m below. Slightly lower down, there is C fault, with a length of 18km, coming up from under the sea. It has the capability of causing an earthquake of magnitude 6.9.

JAEA has admitted the existence of these active faults, but they still deny the danger, saying "there will be no quakes stronger than the assumed level of 760 Gal." Unlike conventional nuclear plants using water, Monju uses 500 degree metallic sodium as a cooling agent. Because of the high temperature, it is structurally necessary to make the pipes thinner, which makes them highly vulnerable to the big power of an earthquake. The authorities who have claimed that 760 Gal of quake resistance is good enough for an earthquake directly below Monju are the same authorities who bear responsibility of the Fukushima meltdown, so you know what that means.

An FBR is hard to operate, easy to run out of control, and a place where a sodium fire can easily break out. Moreover, it is packed full of the most toxic substance on earth, plutonium. All of the countries with the most advanced nuclear technology, including the U.S., U.K. Germany, France, and Russia, have abandoned it. Japan is a follower country in terms of FBR. But only 3 months after Monju started operation in 1995, a sodium fire broke out and operations were halted. It was started up again for the first time 14 years and 5 months later, in 2010; again it was just three months after that that it once again suffered a disastrous accident, and there is no telling when or if it can be repaired. Given this situation, the public should not just sit back and watch, but raise an angry voice, and ask why this useless and dangerous contraption should be sitting there on top of an earthquake fault and burning up 3 trillion yen of the taxpayers' money. Among the people who eat off that money are the puppet-scholars calling themselves nuclear experts who appeared on TV after the Fukushima meltdown to talk their nonsense.

Other than Monju, a grave issue common to the plants on Wakasa Gulf is that there are many aged reactors. More than 40 years have passed since the startup of the following plants: JAEA, Tsuruga, KEPCO, and Mihama. Also, there are 7 reactors that were built in the 70s and 3 reactors that were built in the 80s. On the other hand, only 3 were completed in the 90s, including Monju. Since they are built based on outdated seismic standards, even though seismic adequacy is nominally improved by reinforcement work, it is difficult to believe that all those reactors would demonstrate their aseismic resistance as planned when an earthquake actually happens. If a major earthquake were to shake the whole Wakasa gulf, there's a good chance that all 24 reactors could be destroyed - a waking nightmare.

Shimane Nuclear Power Plant - Local Residents' Anxiety Repulsed by Government and Courts

The Shimane Plant is located on the coast of the Japan Sea at Matsue City in Shimane Prefecture (population about 193,000) and is run by Chugoku Electric Power Company (Energia). Its #1 reactor started to operate in 1974. Currently, there are 2 reactors, and, overriding opposition, they plan to begin operating reactor #3 by the end of 2011. The location used to be called Kashima-cho, Yatsuka-gun before it was consolidated into Matsue City. Now it is the only plant in the prefectural capital, and many people live in front of it.

Major earthquakes in this area include the 2000 West Tottori Earthquake (magnitude 7.3). This was a *shindo* 6 inland earthquake; the maximum acceleration was recorded at Tottori's Hino-cho, at 1,482 Gal. This earthquake, as its name implies, took place in western Tottori Prefecture south of Matsue City. But if things had happened slightly differently, it could have hit the Shimane plant.

Energia had insisted, "there is no active fault near Shimane Plant." However, this is an area with underground energy that could cause earthquakes as big as the West Tottori Earthquake. In fact, as a result of study performed by a group of students lead by Professor Nakada Taka of Hiroshima Institute of Technology, Shinji Fault was found extending east to west, and passing just 2.5km south of the plant. Energia finally acknowledged its existence.

However, the issue is its evaluation. In the beginning, Energia said "the length of Shinji fault is 8km." As the investigation went on, they corrected it to "10km" and in 2008, "it is estimated to be 22km." The longer the fault, the bigger the expected scale of the earthquake. Naturally, appropriate countermeasures are necessary. However, without any evidence, Energia continued to state "Safety is secured even though the active fault is more than 20km". NISA and NSC approved this view. To put it simply, in Japan, any statements from electric power companies will be accepted and approved by the "government puppet" scholars. The situation is equivalent to one in which reactors are operated without any inspection or screening whatsoever.

Against this, Professor Nakada protested, "Investigation shows that the fault is longer than 30km. Energia's research is not sufficient." He also presented his evidence at the Shimane Plant case at which local residents were seeking an injunction against the plant's operation. In May 2010, however, the Matsue District Court ruled in favor of Energia, stating that their aseismic adequacy evaluation was appropriate. The voice of people seeking safety came up against administrative and judicial walls. It is not possible to find a safe nuclear power plant in Japan.

Ehime: Ikata Nuclear Power Plant- Built Above the Largest Active Fault in Japan, the "Median Tectonic Line"

Sadamisaki Peninsula stretches out narrowly into Bungo Channel from Shikoku, in the direction of Kyushu. The Ikata Plant is in Ikata-cho (population about 12,000), which is located on the north side of the peninsula. The Plant is run by Shikoku Electric Power Company (YONDEN). Operation of #1 reactor started in 1977. Currently, three reactors are in operation. Regarding the environment where the plant is located, we need to know that the plant is built above the largest active fault in Japan, the Median Tectonic Line, which was formed by the same tectonic activity that formed the mountains of the island of Japan. Sadamisaki Peninsula itself was in turn formed by the activity of the

Median Tectonic Line. Thus it is natural that there are many active faults in the surrounding area capable of causing earthquakes.

In fact, based on the undersea acoustic wave research of the coast of Iyonada conducted by Professor (of Geology) Okamura Makoto of Kochi University in 1996, there is a report that two seabed faults were found merely 6km off the coast from the plant. They are separated, one to the east and the other to the west, but if they were to move together the total length would be 55 km, so an earthquake with a magnitude of 7.6 could be expected.

Moreover, there are traces that indicate that major earthquakes have occurred here on approximately a 2000-year cycle, and that more than 2000 years have passed since the most recent one. It has been pointed out that if this earthquake should come the Ikata Plant, which would be close to the seismic center, could experience shocks of *shindo* 7 (the highest level on the Japanese earthquake scale) and a maximum acceleration of 1,000 Gal. In 2003, the Earthquake Research Committee of the government released a long-term evaluation that stated, "There is a possibility that 130km of fault near Sadamisaki Peninsula could move. If that happens, the earthquake could be at the magnitude 8 level."

On the other hand, what level of earthquake can the Ikata Plant tolerate? In fact, at Ikata Plant, until the construction of reactor #2, the estimated maximum quake was as low as 200 Gal, based on the belief that the region has few earthquakes. Only 200 Gal – while earthquakes are occurring all over Japan of 2000 Gal or more! After the Great Hanshin Earthquake occurred nearby it was raised to 450 Gal, and after the 2006 Aseismic Code Revision it was raised to its present 570 Gal. But this can't be considered a figure that takes seriously the possibility of a major earthquake on the Median Tectonic Line.

The Great Hanshin Earthquake has also been called the Great Hanshin/ Awaji Earthquake. As this name implies, it was caused by the movement of the Nojima Fault near Awaji Island.   This fault lies on a line that extends northeast from the Median Tectonic Line and through Tokushima Prefecture of Shikoku.   From the fact that that the Nojima Fault has moved, we can infer that that the Median Tectonic Line, which is connected to that fault, is soon to follow.

In Ehime Prefecture, plans or manuals instructing companies on disaster prevention, taking into consideration a possible earthquake caused by the active fault off the coast of Iyonada, assume a possible magnitude of 7.8. This is twice the figure that YONDEN is using.

When managing facilities which could be severely damaged in time of disaster, it makes sense to have a margin of safety.   In this sense Ehime Prefecture's approach stands to reason.   However, NISA and NSC judged YONDEN's safety standard to be adequate, and gave them the go-sign. What kind of government is this?!   And who will protect the people's lives?

Saga: Genkai Nuclear Power Plant - Supposedly Safe Area But Suddenly Hit by Quake of M7.0

Genkai Plant is run by Kyushu Electric Power Company (Kyushu Electric).   It is located in Genkaicho, Saga Prefecture (population about 6,000), facing Ikinoshima on the opposite shore.

Reactor #1 started operation in 1975, reactor #2 in 1981.   Both reactors are old, having been in operation for more than 30 years.   In the 90s #3 and #4 reactors were completed, and currently, all four are in operation. In December 2009, #3 became notorious as the first reactor in the country to begin pluthermal operation using plutonium MOX fuel, which triggered a torrent of protests from all over the nation.

In December 2010, based on test pieces taken for the 2009 regular inspection, it was found that the brittle transition temperature of Genkai's #1 reactor (36 years in operation) had reached 98 degrees.   What this little-known term, "brittle transition temperature" tells us is that the thick steel of the reactor, being subjected to neutron bombardment every day, becomes brittle, and can fracture with a drop in temperature. That is, if an accident were to happen and cold water came against the steel, it could shatter the way a glass shatters when it comes in sudden contact with boiling water. A reactor whose steel has become brittle could simply split open, leading to a horrific nuclear accident which so far has never taken place on the face of this earth. People first learned about this phenomenon from an accident when a large ship broke apart.

Kyushu Electric, however, claiming that this aging reactor is safe, nonchalantly keeps it in operation.

Then what are the earthquakes that could cause such an accident? The area around the Genkai Plant has been known as an area where big earthquakes hardly happen. Historically, even earthquakes at the magnitude 6 level have been recorded only twice: the Tsukushi Earthquake of 689 and the Itoshima Earthquake of 1898. Looking at this data, the area seems to experience a major earthquake literally "once in a thousand years."

On 20 March 2005, however, there was a magnitude 7.0 earthquake offshore from western Fukuoka Prefecture. Its seismic center, at the bottom of the sea, was only 40km from the Genkai Plant as the crow flies. It had not been known that there was an undersea fault in that area. In a place where it was believed there were no faults, in fact a fault was hiding, and we now know that it is a fault capable of producing a major earthquake on a scale never before recorded in this region.

If you come to think about it, it was in the 4th century when writing was imported to our country from China. Thus, earthquake records in Japan go back only 1500 years or so. Also, in less populated areas, writing came in still later. We have to remember that if we rely only on past written data, we might fail to grasp the unbelievable destructive force of nature, a failure we might deeply regret later.

With this in mind, we should pay full attention to the fault complex of off the coast of Itoshima Peninsula, only about 20km from the Genkai Plant. If we relax our vigilance because we have no written data showing major fault activities in the past, when an earthquake actually does come in that spot we will have much to answer for. To repeat the maxim given us by Terada Torahiko: "Natural disasters occur when you aren't thinking about them." This is a seabed fault 23km in length. Kyushu Electric estimates that if this fault moves, an earthquake of magnitude 7.1 class could occur. I think, however, that this is an underestimation. The aseismic adequacy of the Genkai Plant is 540 Gal, lower than the average of other reactors in the country even after the revision of the aseismic code in 2006. What are the electric companies thinking while 2,000 Gal class earthquakes are frequent in the country these days? Thinking about it makes me uneasy. Moreover, I hardly understand the mentality of Kyushu Electric, which is running a pluthermal operation, the riskiest operation in existence. If reactor #3 should be involved in a major accident, the whole Kyushu region would be exposed to plutonium and destroyed.

Kagoshima: Sendai Nuclear Power Plant - Kept in Operation during an M6.2 Earthquake

The Sendai Nuclear Power Plant is located on the western coast of Kagoshima Prefecture facing the East China Sea. The plant is operated by Kyushu Electric. Operation of #1 reactor started in 1984; #2 reactor started operation in the following year. The latest plan, decided on in 2009, is to build a third, mega-reactor of 1,590,000 kW capacity, which would make it the biggest in Japan, or for that matter in the world (meaning it would contain the most radioactivity).

Sendai plant is located in Satsuma, Sendai City (population about 100,000), at the end point of the Median Tectonic Line, the biggest active fault in Japan. In addition, there is another huge fault, the Butsuzo Tectonic Line running from Inubosaki, Chiba Prefecture, parallel with the Median Tectonic Line. Those two tectonic lines lead to the Nansei Islands farther to the west to form the Pacific Ring of Fire. Why were nuclear plants built in such a location?

Especially along the Median Tectonic Line, after the 1995 Great Hanshin Earthquake, there has been much earthquake and volcanic activity. The Sendai Earthquake, which occurred in March, 1997, is believed to have been a part of that series. It was a medium sized earthquake, measured at first at magnitude 6.2, a figure which was later revised to 6.8. In Sendai City, its force was measured at *shindo* 5 (on the Japanese scale), and roads and communication networks were severely damaged. Nevertheless, Kyushu Electric did not shut down the operation of the Sendai plant but kept it running, which greatly intensified the distrust of the local people. What should be noted here though, is that this was the first time that a relatively large earthquake hit a nuclear plant in Japan.

Almost half a century has passed since Japan launched nuclear power plant construction. Fortunately, it fell in a period of seismic quiescence in the Japanese archipelago. And it was the period when the Japanese were vigorously pushing high economic growth. Now, however, the archipelago has entered a period of large scale movements of the earth, and major earthquakes have started to occur close to nuclear plants. The great destruction at the Kashiwazaki-kariwa Plant (2007), and, at the Fukushima Daiichi Plant, triggered by The Great Tohoku Earthquake, the first genpatsu shinsai syndrome.... In that sense, the Sendai Plant is an epoch making plant, the first to give warning of the danger.

If that was the first warning of the Fukushima Daiichi Plant meltdown disaster, then what was the last? Immediately before the Great Tohoku Earthquake, that is, on 26 January, 2011, the mountain Shinmoedake in the Kirishima Range began a major eruption. This eruption is linked to the fact that the eruption of Sakurajima, in Kagoshima Prefecture began to intensify 2 years ago, in 2009. In the Taisho era, the Ohachi volcano in the Kirishima Range erupted and four days later, on 12 January 1914, Sakurajima erupted. The eruption was so big that it connected the island of Sakurajima to the nearby Oosumi Peninsula. Over the last two years, when I give talks I have been warning people, "The Great Kanto Earthquake occurred 9 years after the big eruption of Sakurajima. The next big earthquake is coming soon!" But nobody could think of past incidents in southern Japan in the context of possible events in northern Japan. Sakurajima erupted 548 times in 2009 and 896 times in 2010, which is a record. Considering global plate movement, the activity of Shinmoedake was the final omen presaging a major earthquake off the Pacific Coast.

With that in mind, consider well the location of the Sendai plant. It is in Satsuma Sendai City, within the Hokusatsu Volcanic Cluster which is joined under the earth to the presently erupting Shinmoedake of the Kirishima volcanic cluster. Magma is seamless. Along the Median Tectonic Line, the signs that a major earthquake and a major volcanic eruption are approaching are becoming clear. Nevertheless, Kyushu Electric is assuming that an earthquake caused by nearby active faults would have a magnitude of 7.3 and has set the aseismic standard at 540 Gal. They feel no fear, and have no policy. Unless it is accepted that a magnitude 8 earthquake is impending, and the Sendai Plant is discontinued as soon as possible, life in Kyushu cannot be guaranteed. I appeal to the people of Kagoshima Prefecture: after the accident happens, it will be too late.

I have looked through the "Nuclear Plant Archipelago" from north to south. I cannot suppress my amazement that on such narrow islands, laced with active earthquake faults, and with earthquakes and volcanoes coming one after another, so many nuclear power plants have been built.

Rather than the faults attacking the plants, doesn't it seem as though the plants themselves are intentionally shaking these eerie, hidden faults out of their long slumber?

Along with the hair-trigger situation of nuclear plants built over active faults all over the islands, the rifts that could invite the final catastrophe are all connected inside the globe. What I most strongly hope the readers will see is this global-scale movement. Not the specialized details of seismology; I hope they will come to be able to anticipate these historic movements of the earth.

Concluding Chapter:
Japan's Nuclear Power Policy is Completely Bankrupt

According to the International Atomic Energy Agency (IAEA), as of January 2010 there were 437 nuclear reactors in operation in the world. The USA has the largest number with 104. France is second with 59 and Japan with, as we have seen, 54, is number three.

While after the Chernobyl nuclear accident of 1986 there was a mood around the world to reconsider nuclear energy, more recently with growing concern about global warming and with increasing demand for electricity there has been a "nuclear Renaissance" - - or so the story goes, but this is a great falsehood. If it looks like a developed country like the US or the UK is going to build one or two reactors this is reported with a great clamor, but if you look at the situation of the nuclear power industry as a whole, it is in a state of general decline. In 21st Century, nuclear power plants in the developed countries are reaching the end of their life spans one after another, and we are entering an age in which more and more shut-down reactors are being dismantled.

However here in Japan, we cannot learn that the age of nuclear power is coming to an end by watching television and reading the newspapers. Still harder to believe is the fact that, while after the Fukushima Daiichi accident one would expect a growing number of people deciding that they had had enough, and saying no to nuclear power, opinion polls taken after the accident show that a majority continue to support the status quo. Germany responded to Fukushima by adopting a policy of early shutdown of its nuclear plants; Switzerland responded by putting a freeze on all new construction; the West as a whole is shifting toward a policy of moving away from nuclear power. While this goes on, in Japan public opinion supports the status quo, meaning that Japan should continue as the number three nuclear power. Are the nuclear power industry's boasts and threats – that nuclear power is the only clean energy that can be effective against global warming, that without nuclear power the public's demand for electricity cannot be met – still exercising authority?

The time of the next major earthquake, which is to say, the time of the next genpatsu shinsai syndrome after Fukushima Daiichi, is steadily approaching. Japan's 54 nuclear reactors are in any case approaching the hour of their demise, but if before that happens a major earthquake comes, that will bring the story to its end. I see it as the destruction of the Japanese Archipelago.

In this final chapter I will once again describe the true situation of Japan's nuclear power plants. What to do about them will depend on the will of the Japanese people.

Is Nuclear Power the Trump Card Against Global Warming?

In recent years there seemed to be a nuclear power renaissance. One reason for this has been the adoption by its promoters of the theme of global warming, and their claim that nuclear power is clean energy because it does not produce carbon emissions. Since the Fukushima accident nuclear power discourse, with its false claims of economy and its dismissive attitude towards questions of safety, has undergone a sharp change of direction.

Is nuclear power in fact the clean-energy solution to global warming?

In Chapter 1 I explained the circulation of heat inside a nuclear reactor, and showed that only one third of the heat energy that it produces is transformed into electricity. The remaining two thirds of the energy that remains in the water vapor– that is, twice as much energy as contained in the generated electricity – is disposed of in the sea. In the cooling system, seawater is used to cool the water vapor, whfich condenses again to water and is circulated through the reactor once again. This heated seawater is called "thermal discharge". How much heat does this thermal discharge carry into the sea? The amount is startling.

Before the Fukushima accident, that is, at the end of 2010, Japan's 54 nuclear reactors were producing a total of 49,112,000 kW of electricity. So every day they were throwing away twice that much, approximately 100,000,000 kW of energy, in the form of heat, into the sea.

This means that every day they were pumping into the sea energy equivalent to 100 of the atomic bombs dropped on Hiroshima. The Hiroshima bomb destroyed the city in an instant and ended the lives of some 140,000 people, but when energy 100 times that great is "dropped" into the sea daily, what effect does that have? That it would not be destructive of the ocean's ecology is unimaginable. Before saying that "nuclear power plants supply one third of the demand for electricity", it needs to be said that "twice as much energy as the electricity they produce is used to heat up the sea."

I want to ask, what kind of global warming debate is it that never discusses this fact? In Japan, the number one global warming agent is the nuclear power plants.

After I left the company I was working for, I spent a long time translating medical books. In the 1970s I was translating books depicting the suffering of people whose health was damaged by environmental pollution, and at the same time through an agent was accepting work from industry. At that time I received a request from

TEPCO to translate a 1970s report from the Organization for Economic Co-operation and Development (OECD). In it was the following passage.

"When thermal discharge from nuclear power plants is released into the sea, the heat does not immediately disperse. Rather it concentrates and remains suspended in what are called "hot spots". For this reason it has a very large effect on sea life near the shore. In the shallows, even a difference of two or three degrees can kill fish eggs or young fish."

I translated this English correctly and delivered the manuscript to Tepco. The report of which it was a part was apparently suppressed within the company. To this day it has never appeared.

Moreover, the claim that nuclear power is a cheap form of energy is also untrue. Nuclear power plants are located far from the users of the electricity, so they require extraordinarily long transmission systems (In 1964 the Japanese Atomic Energy Commission (JAEC) stipulated that "Dangerous nuclear power plants must not be located in heavily populatedareas"). The nuclear power plants that deliver electricity to the capital are the Fukushima Daiichi and Daini reactors, Niigata Prefecture's Kashiwazaki-Kariwa reactor, and Ibaraki Prefecture's Tokai Daini reactor. The 14 nuclear power plants sending electricity to the Kansai (Kyoto, Osaka, Kobe) area are lined up along the faraway shore of the Japan Sea at Wakasa Bay, in Fukui Prefecture. When you take into account the transmission systems connecting the power plants with the metropolitan areas they serve, you cannot call it an inexpensive source of electricity.

I treated the question of global warming in relation to Japan's energy situation in my *Nisankatanso Ondankasetsu no Hokai* (*The Collapse of Carbon Dioxide-Global Warming Theory*, 2010, Shueisha Shinsho); readers who still have doubts are invited to consult that.

Without Nuclear Power, Will There Be Blackouts?

After the Fukushima Daiichi accident, TEPCO has carried out planned blackouts, and the Kan Naoto administration, "in order to avoid a major blackout due to electricity shortages in the summer months" is considering enacting measures enforcing limits on electricity consumption for the first time since the oil shock of 1974. This deep-seated "blackout fear" held by so many seems to be grounded in the idea that we must continue gingerly to maintain the nuclear power industry, which advertises itself as providing one-third of the country's electricity. What I see in the opinion polls is the attitude, I don't like living with nuclear power plants, but without them there is no way to get the electricity, so there's nothing to be done because like they say, you can't exchange your back for your belly.

This is a huge misunderstanding that must be corrected.

Look at Figure 18. It is based on the research of Fujita Yuko, formerly Associate Professor at Keio University, and shows the changes by year in the generating capacity of Japan's main sources of electrical power, compared to the total amount of electricity demand. The bars show the total production capacity for each year by fossil fuel, water, and nuclear power plants respectively. The jagged line shows peak demand, that is, the demand for electricity between 2 and 3 PM on the hottest days of summer.

You can see that even without the nuclear energy shown at the top of the bar graph, the electrical capacity of the fossil fuel and water powered generators is sufficient. Moreover, the highest recorded peak demand was in 2001, and has never been surpassed in the almost ten years since then. Rather, with the economic downturn, (although the graph doesn't show this) demand for electricity fell in 2008 and 2009.

From whence, then, comes the misunderstanding that nuclear power plants supply one third of the country's electricity, and that without them there would be blackouts?. The answer sounds like a joke, but it is true: it is that while the Japan has a very large capability for generating electricity from natural gas, these facilities have been intentionally kept operating at

Figure 18　Trend of capacity and maximum electric power of power generation facility

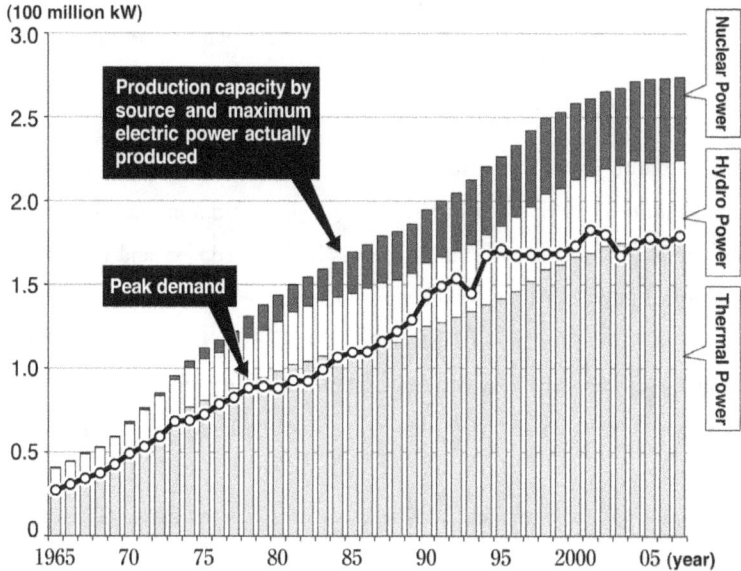

(100 million kW)

The highest electrical comsumption has never exceeded the capacity of hydro plus fossil fuel combined, so there is no danger of blackouts even without nuclear power.
"Enerugi Keizai Tokei Yoran" (Energy / Economic Statistics Directory) (1994-2009 edition) By : Yukou Fujita

only 50-60% of capacity. Among the major sources of electricity used in the advanced countries, natural gas is the cleanest. Then there are the petroleum powered plants; amazingly they are operating at only 10 to 20% of their capacity. (This figure may sound unbelievable, but since the 1970's Oil Shock, most of the developed countries have a policy of reducing oil consumption as far as possible. Japan's fossil fuel power plants use mainly coal and natural gas.) The idea that without nuclear power there would be blackouts is nonsense.

The reason TEPCO carried out intentional blackouts after the earthquake is that the fossil fuel reactors in the region also suffered temporary damage. No doubt there was also difficulty delivering fuel. But repairing fossil fuel power plants is nowhere near as difficult at repairing nuclear power plants. It's just a matter of replacing damaged parts. Once repair work begins, it doesn't take long before the plant is operating again. And once the fossil fuel plants are back on line, electricity demand is no problem.

After its nuclear plants were so badly damaged, TEPCO should have put its natural gas and petroleum plants into full operation, but it did not. Rather it carried out intentional blackouts, bringing confusion to the metropolitan area and bringing losses both to industry and to private citizens. In this it did not fulfill its responsibility as an electricity provider, and revealed a fundamental problem. And now we hear everywhere language fanning the fear of summertime blackouts, but this is only a false rumor being spread by people who know nothing of electrical power generation.

A natural gas power plant can be built in a few months. This was made clear in an article appearing the April 6, 2011 edition of *Gasu Energii Shinbun* (*Gas Energy News*) by Ishii Akira, head of the Energy and Environment Research Center, titled "After Fukushima, the Age of Natural Gas". In this article, Ishii explains

Japan's energy situation from the standpoint of a professional. The Fukushima nuclear power plant accident took place on March 11. Why didn't TEPCO begin immediately to take action to ensure that there would be no electrical shortage? If they couldn't get it done in time, why did they not immediately ask the world's largest manufacturer of natural gas power plants, America's General Electric (GE) to do it for them? An electric company that can't supply electricity to the public has no right to be called an electric company.

Nuclear power supporters will argue that the supply of natural gas is limited. But this too is the outdated opinion of one who does not know the energy industry. As Ishii Akira pointed out in an article of Feb 2, 2011 in Gas Energy News, new sources of natural gas are being discovered one after another all over the world. In the Mediterranean Sea, offshore from Madagascar, under the sea to the east of India, on the continental shelf in northwestern Australia, in Brazil, in Turkmenistan – in the ten years up to 2009 the world's known supply of underground deposits has increased by close to 30%. In addition to this natural gas supply, new, so-called non-traditional gases such as coal bed methane, tight sand gas, shale gas, and methane hydrate are being developed one after another. According to Japan Oil, Gas and Metals National Corporation (which is dedicated to locating natural resources for Japan) the underground reserves of these new forms of natural gas total more than 922 trillion cubic meters, more than five times the reserves of traditional natural gas. No doubt there will be future discoveries one after another, so I would say that we have enough gas reserves alone to last well over 200 years.

Another important factor is the Japanese government's ignorant lack of policy over the years. The Japanese public doesn't know this, but the electric companies are not the only organizations in the country that can supply electricity. Assistant Vice-Minister Nishiyama Hidehiko of NISA sought to intimidate the public by saying that "Without nuclear power plants Japan cannot survive. The alternative to nuclear power plants is blackouts," but it is because people like this who know nothing about electrical power are the movers of government policy that the public as a whole is kept in ignorance. That a person so incompetent should be serving as Assistant Vice-Minister is enough to leave one speechless.

If all of the companies in Japan that generate power on their own – the so-called Independent Power Producers (IPPs) – were put to full use, there would be absolutely no danger of blackouts. According to a December, 1997 report prepared by an agency of METI, after the system of bidding for wholesale electricity was introduced in 1995 it became clear that the IPPs taken together have a latent capacity for producing

21.35 to 34.95 MW (megawatt = one million watts) of electricity, and that if the system is improved this could be raised to as much from 38 to 52 MW. (These figures are not the total amount they generate, but the total amount they would be able to sell, not counting the electricity they need to operate their own factories.)

"Improving the system" simply means taking away from the electric companies their monopoly of the electric transmission systems, by which they prevent other electrical producers from using them. Electrical generation and electrical transmission should be separated, and the state should manage the transmission systems in the public interest. If the IPPs can produce 52 MW of electricity, then even if all the nuclear power plants, whose total capacity is 50 MW, went down at once, there should be no problem; wouldn't you think the government would notice that? And with all these new wholesale providers entering the market we could expect the price to go down, so the public, in addition to being freed from the fear of nuclear power, would get a double benefit. The industries I refer to, are for example Nippon Steel, Ebara Corp., Showa Denko, Tomen Electronics, Hitachi Zosen, Nippon Petroleum Refining, Hitachi, Kobe Steel, Idemitsu Kosan, Nippon Paper, Kawasaki Steel (now JFE Steel), Cosmo Oil, Ube Industries, Tokyo Gas – the list goes on and on. This massive collection of Japan's representative major industries, taken together, has the capacity to generate a huge amount of electricity. If the government were to direct them to generate electricity for public use, the problem would be solved overnight. If I were prime minister, that is the policy I would take. Japan is not a pre-industrial country; blackouts are not an issue.

In Japan, nuclear power plants are absolutely unnecessary. It's time to bring to an end this ignorant, amateurish debate in the mass media about blackouts. What the people need for their lives and their industry is not nuclear plants and radiation, but electricity. Wouldn't it be a fine thing if we could get our electricity without depending on these electric companies that keep on propagandizing for nuclear plants, that control the TV channels, and that spread rampant corruption among our politicians and our bureaucrats?

What is needed now to save our nation is a change in the consciousness of each member of the public. It means, knowing the facts.

That even without nuclear power plants electrical demand would not be affected has been demonstrated in practice. This demonstration took place on April 15, 2003. At that time, TEPCO was trying to cover up a major failure from the year before, its falsification of the data was discovered, and it was forced to shut down the Fukushima Daiichi and Daini Power Plants and the Kashiwazaki-Kariwa Power Plant – a total of 17 reactors in all. Then, as now, there were dire predictions of a midsummer electricity crisis, but in fact when summer came there were no blackouts.

To build and maintain nuclear power plants for the purpose of meeting the peak demand, which is just a brief period in the afternoon in midsummer, is in the first place the height of absurdity. The peak demand period is mainly a result of the activity of businesses, local governments and schools, not of individual homes.

If a system were established whereby the price of electricity were higher at the peak periods, businesses, which are always on the lookout for ways to save money, would begin to conserve electricity during those times, and the consumption rate would quickly go down. Many other enlightened methods for regulating the electric industry are possible. Electric power stations and electricity-producing facilities should be divided up, localized, and made as small as possible, and to achieve this, what is needed is for the people to raise their voices and demand an end to the electric companies' monopoly over electrical production, and a separation between the generation and the transmission of electricity.

Japan's Nuclear Power Plants at a Dead End

Fukushima Daiichi Power Plant is continuing to release radioactivity into the atmosphere and into the sea.

As I write this manuscript, people have been coming to me and asking in desperate voices, "What should we do now?" Both the government and TEPCO have said that the reactors will be shut down. But can they disassemble and get rid of them? Can they bury them under concrete and turn them into sarcophaguses, as was done with Chernobyl's No. 4 reactor?

I don't know what should be done. It was because I knew that if it came to this nothing could be done that I opposed nuclear power. If I had thought there was something that could be done, I would not have opposed it. And of course neither TEPCO nor the Japanese government has made even a hypothetical plan for what to do with a nuclear reactor that has begun to leak.

However I do think I understand what is going to happen now. I base the following scenario on the optimistic premise that the leakage at Fukushima Daiichi is somehow put under control, and its reactors are finally shut down. At that point it will become publicly clear that all other nuclear power plants in Japan are in the same terminally wretched condition, which will lead them to the same fate. And Japan's nuclear power society will come to a crashing end.

If the nuclear power plants continue in operation, they will soon find themselves in a street with no exit, and will have to be shut down.

That is the last thing I want to write about.

The Final Disposition of Meltdown

About two hours by car from downtown Aomori City, in a hilly area at the root of Shimokita Peninsula, there is a large factory occupying about 3,800,000 square meters of land.

At the time of the Great Tohoku Earthquake, as I sat before the television shocked by the force of the trembling and watching the disaster unfold, the image of genpatsu shinsai syndrome that came to me first was not at Fukushima Daiichi, but at this complex. Here at Rokkasho Village is located the massive Rokkasho Reprocessing Plant, where spent nuclear fuel is supposed to be recycled.

In the thirteen years between 1998 and 2010, the spent nuclear fuel from all of Japan's 54 nuclear power plants has been brought here, incredibly totaling now some 2827 tons (uranium weight equivalency, which means the weight of the uranium only, not counting the oxide, in the uranium oxide). If this facility, like Fukushima Daiichi, suffered a failure of electricity or damage to the fuel storage pool resulting in an explosion, presumably the result would be something beyond imagination, equivalent to the simultaneous explosion of some 100 nuclear reactors. It would be the end of the Japanese Archipelago, or rather, the end of the world.

Before the 3/11 earthquake came, No. 4 reactor at Fukushima Daiichi had been shut down. Nevertheless, four days later on the 15th the cooling system on the spent fuel storage pool failed, causing a hydrogen explosion which blew off the roof of the building. When that happened, a new fear entered my mind. What this explosion taught us is that the greatest danger in Japan is the huge storage pool for spent nuclear fuel at the Rokkasho Reprocessing Plant. This storage pool is constructed of concrete, surfaced on the outside by many metal plates welded together. A major earthquake could easily split the welds. But a greater danger lies in the possibility of an electrical failure causing the cooling system to shut down. And in fact, just under a month after the 3/11 earthquake, on April 7, the strongest of the aftershocks caused a power failure in the four prefectures of Iwate, Aomori, Yamagata and Akita. During that time the outside supply of electricity at the Rokkasho Reprocessing Plant also failed, and the cooling system for spent fuel and for radioactive liquid waste was powered by emergency generators. That is, we were just one small step away from the abyss. The Japanese mass media and the public seemed untroubled by this. As shown on graph #3, the Rokkasho plant is 55 m. above sea level and 5 km. from the shore, so it is build on the assumption that no possible tsunami can reach it. Do you believe that no tsunami can climb a gentle slope to a height of 55 meters? After the Great Tohoku Earthquake we all saw on television what a tsunami can do, overwhelming all barriers, gobbling up all before it. And in fact this tsunami did, in some places, reach as far as six kilometers inland.

Construction on this huge chemical plant, designed as described in Chapter 2 to extract, from the spent fuel of all the country's nuclear power plants, plutonium and uranium 235, was begun in 1993. The plutonium and uranium obtained by this process, which they call reprocessing, can then be used again as fuel; thus the purpose is to achieve a nuclear fuel cycle.

However this process is originally the method for producing weapons-grade plutonium, and is dangerous in the extreme. Without going into the details, it involves the use of extremely dangerous chemicals. The technology itself was imported from the French company Cogema, later Areva. This is the company that shamelessly sent a delegation to Japan, headed by its CEO Anne Lauvergeon, allegedly to help clean up the pollution after the Fukushima Daiichi accident. Lauvergeon was a favorite of the late French President Mitterrand, and is a kingpin of the nuclear power industry. From this source Japan imported, and began operations with, the still only half-developed technology of reprocessing, and it has been a stop-and-go operation ever since. To this day they still cannot even complete a test run, and the process seems to be approaching a dead end. There appears to be little prospect that the technology will ever become operational.

As a result, all the spent fuel from all the nuclear power plants has been collected at Rokkasho where, almost none of it reprocessed, it fills the storage tanks to near overflowing. The tanks have a storage capacity of 3000 tons, and they already contain 2827 tons, which means they have room for just 173 tons more.

Japan's nuclear power plants produce between 900 and 1000 tons of spent fuel each year. This means that this waste can no longer be carried to Rokkasho, but will have to be kept on the premises of each plant. That is, the "nuclear garbage" produced year after year by these plants has no place where it can be disposed of, so there will be no choice but to store it where it is produced.

However, as you can see from Figure 19 (page 162), the amount of time remaining during which spent fuel can be stored at each plant is, for most of the plants, less than 10 years, and is on the average 8.1 years. So unless some new place for disposing of this nuclear garbage is found, Japan's nuclear power plants one after another will find themselves in a situation in which they have to stop operations. Put in order, here is the situation.

Nuclear power plants burn uranium to generate electricity. When they do this, it produces spent fuel that contains plutonium and highly radioactive nuclear waste.

The plan was to transport this to Rokkasho Village, where it would be reprocessed so that plutonium and other substances would be extracted from it, and a nuclear fuel cycle would be achieved. However at Rokkasho Plant all tests runs so far have failed before being completed, and so the spent fuel storage pools are full.

If the Rokkasho plant ever does succeed in operating so as to reprocess spent fuel, this will lower the levels in the storage pool, but it will also produce as a byproduct large amounts of highly radioactive liquid waste.

Figure 19
The amount of time remaining during which spent nuclear fuel can be stored at nuclear power plant

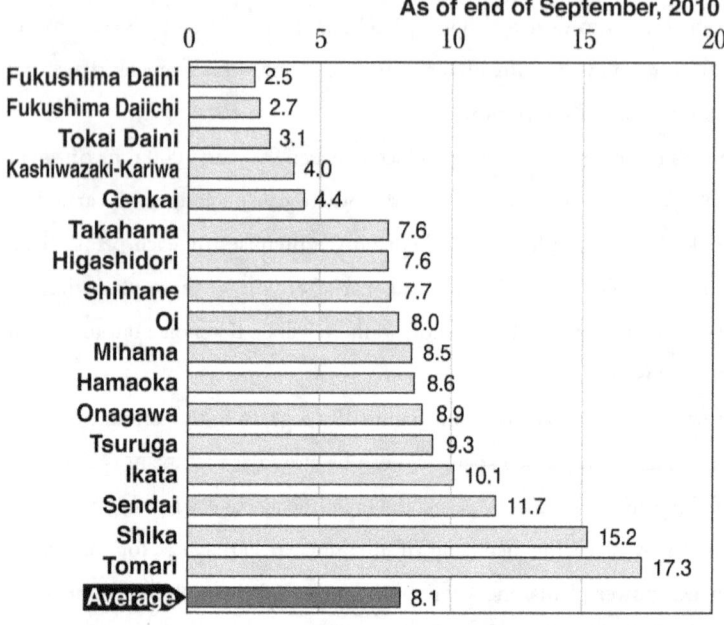

As of end of September, 2010

As of 2010, plant operation should be stopped once in about 13 months for 3 month inspection. 13 months + 3 months = 16 months. So it is assumed that fuel is replaced every 1.33 year. This is the way to calculate the amount of time.

If, as a result of an electrical failure, or of earthquake-caused damage to the circulating pipes, the cooling system for this radioactive liquid waste were to fail, this could cause an explosion that would destroy all of northeast Japan and Hokkaido. Thus the only thing to be done with it is to seal it up in explosion-proof glass and bury it 300 m. or more underground.

Thus what is needed is a final resting place other than Rokkasho Village for this spent fuel, but in all of Japan there is no place where such dangerous waste would be accepted. Thus the nuclear fuel reprocessing plan has come to a stalemate.

It will only be a few more years before the storage space for the nuclear waste from Japan's nuclear power plants will have run out. With no place to store spent fuel, all those plants will have to stop operations.

And what will happen in the end to the Fukushima Daiichi Power Plant? Given the intelligence level of Japan's nuclear power industry, there is a danger that radiation will go on dribbling out for years to come, pseudo-scholars will go on paralyzing the public with their pseudo-scientific assurances of safety, and more and more people will become radiation victims. But even if we assume the best case, where leakage is entirely sealed off, there will remain the question of where in Japan to dispose of the melted fuel rods, the glob of radioactive material produced by the meltdown. Does any such place exist? At that point the fact that Japan's nuclear power strategy is bankrupt will become clear to the public, and that strategy will be borne down under the double burden of the cleanup at the site and the problem of disposing of the remaining radioactive material.

The spent fuel produced by nuclear power plants contains large amounts of plutonium, which has a half life of 24,000 years. That means it must be kept distant from the environment in which people live for hundreds of thousands, or even millions, of years. Can we say it is the responsibility of the children, the people of the next generation, to take care of this? Here on the Japanese Archipelago, where the next massive earthquake might come tomorrow, is there anyone who can imagine that far into the distant future?

After our generation is gone, who is going to manage those waste sites? Has our generation seriously considered what a crime we are committing against the innocent future generations? Surely it is safe to make a flat statement that, sometime during the period when this waste is being stored, a major earthquake is bound to make a direct hit on one of the nuclear power plants.

For human beings and nuclear power to live together is, from the standpoint of earth science, an impossibility.

The Japanese public and business circles must by now have realized the size of the blow suffered by the economy, once the stigma of Radiation Polluted Nation has attached to it. Like frogs at the bottom of a well, talkers on television have tried to keep in the favor of the electric companies by issuing their insipid "No danger! No danger!" statements one after another. When politicians come from abroad with the intention of helping, the result is no more than a revolting solidarity among politicians and a string of falsehoods tossed off to the media. If the Japanese people continue to believe this kind of low-level news reporting and keep their mouths shut, the world will pass on by and leave the country and its industry behind and isolated. If the people don't come to grips with the seriousness of the danger of the ongoing nuclear disaster and show the decisiveness to put an end to the country's nuclear power program immediately, the world will have no reason to believe in the Japanese intelligence.

We must not lose hope. If we rid ourselves of nuclear power plants, we can open up a new future for our country.

Are we going to choose to go on living in the dark, anxiety-ridden society of nuclear power plants?

Shouldn't we work together to build a society where our minds and bodies are liberated from that fear, and living is that much easier?

I promise you: a life with the danger neither of nuclear radiation nor of blackouts can easily be achieved.